Rocks, Gems, and Minerals

Third Edition

Garret Romaine

FALCONGUIDES

GUILFORD, CONNECTICUT

To my uncle Doug, who never turned down a
chance to head out to the field.

FALCONGUIDES®

An imprint of The Rowman & Littlefield Publishing Group, Inc.
4501 Forbes Blvd., Ste. 200
Lanham, MD 20706
www.rowman.com
Falcon and FalconGuides are registered trademarks and Make Adventure
Your Story is a trademark of The Rowman & Littlefield Publishing Group, Inc.

Distributed by NATIONAL BOOK NETWORK

British Library Cataloguing in Publication Information available

Library of Congress Cataloging-in-Publication Data available

ISBN 978-1-4930-4686-7 (paper : alk. paper)
ISBN 978-1-4930-4687-4 (electronic)

♾™ The paper used in this publication meets the minimum requirements
of American National Standard for Information Sciences—Permanence of
Paper for Printed Library Materials, ANSI/NISO Z39.48-1992.

Contents

Overview

About Geology

The term *geology* is a combination of two Greek expressions: "Geo" refers to the Earth, and "logos" refers to the logic and language used to explain your observations. So think of geology as a way to organize and explain the Earth processes that we see all around us. For some of the strange shapes we see, an educated explanation would be great. Most of what we know is good guesswork, based on lab experiments and inferences that take many detailed drawings to explain and a lifetime to understand. Fortunately, the more you see, the better things fall into place.

Geology is a young science, dating to 1815 if you start with William Smith's first geology map. Still, we know the ancient Greeks, such as Pliny the Elder, described various rocks and minerals, and many other scholars documented the metal mines of their times. Later, some of the smartest and most educated scientists laid down the basics, such as Nicolas Steno (1638–1686), who observed that most of the time, the rocks at the bottom of a cliff are older than the rocks at the top. He called that the Law of Superposition, and with a few other theories, Steno could explain how fossil seashells ended up on mountaintops. He also developed the Principle of Original Horizontality, which states that sedimentary rocks are usually deposited flat, although there can be local pinching, advancing, and other variations. Later, James Hutton presented *Theory of the Earth* in 1785, and Sir Charles Lyell wrote *Principles of Geology* in 1830. Arguments soon arose over the question of whether geology happened in slow, methodical processes or in short, catastrophic bursts. Eventually distinguished thinkers realized the answer was actually "both."

There are two key points to consider when trying to understand geology: time and entropy.

- **Time.** Earth is a very young planet and thus still very active. Even though scientists have measured Earth at 4.54 billion years old, that's young in the context of a 20-billion-year-old universe. Given enough time, a lot can happen. The forces that boil up from the Earth's magnetic core are a long way from burning out, and they are relentless. Some activities happen quickly, like tsunamis, and we have the video. Other forces take millions of years, leaving clues like all the mica flakes lined up in a schist. Good field observers can identify the obvious signs of things that seemed to happen before and apply those signs to the present and future.

- **Entropy.** Things fall apart all the time. Stuff happens. Storms rearrange coastlines and rework river channels. Earthquakes, volcanoes, windstorms, and floods all move mountains and leave scars that "heal." A rock balanced precariously atop another rock will not remain for long; eventually, it will shake loose. The Earth is very efficient at recycling all that surface mayhem, hiding many clues. Mountains rise and then get ground down under glaciers and unrelenting rain. Tight chemical bonds that hold atoms together eventually weaken, thanks to water, heat, pressure, and time. Oxygen in the air constantly rusts iron and

Agates and petrified wood are often the first additions to a rockhound's collection.

dissolves minerals. Those forces are always at work and are easy to predict but sometimes hard to imagine. Try to picture the Mississippi River under flooding conditions that happen once every 1 million years. That's mayhem on a continental scale. Now imagine the resulting gravel bars as the river recedes from flood stage and think about the possibility of being the first rockhound to check for agates on those newly stirred deposits.

Given enough time, almost anything can happen, and it usually does. There is a lot of math, chemistry, physics, biology, and just general science involved in sorting out what's going on in the field. But you're mostly interested in what you can see and collect, so read on.

Think in Series

We don't get many absolutes in nature, but numbers, such as percentages of minerals present, help when thinking about crystal compositions. Just as there are probably no two snowflakes that are exactly alike, most granites differ in some way. Some basalts may have more iron and magnesium present, and some may have more feldspar. Some may have more magnesium and some more iron. You can't exactly tell without expensive equipment. And usually it doesn't really matter to that many decimals if you are trying to figure out if you have a calcite vein or a quartz vein. You don't need a PhD in structural geology to dig out a seam of agate. You do need a hard hat, however.

In addition, rocks that start out as one thing can turn into several different rocks after billions of years. Consider the metamorphic progression below:

Increasing metamorphism ⟶

Mudstone	Shale	Slate	Phyllite	Schist	Gneiss

Each stage along the way represents a different setting for heat and pressure that altered the preexisting rock. The boundaries may even be fuzzy or under dispute. Memorizing the series isn't the important thing—you just need to know it exists for right now.

Then there's the series just for schist:

Increasing metamorphism ⟶

Chlorite	Biotite	Garnet	Staurolite	Kyanite	Sillimanite

Again, there are specific temperatures and other conditions scientists measured in lab experiments. There are reaction series, ranges of sediment sizes, and other different ways to explain rocks, minerals, and Earth processes.

Field Studies

One big complication in identifying rocks and minerals is the tendency of Earth's atmosphere to oxidize everything. That same oxygen we need to breathe also wreaks havoc on fresh material. Oxygen ions are always looking to hook up with another ion, preferably a metal. You've seen how a freshly sharpened knife blade starts to rust in the rain. That oxidation is also at work on cliff faces and boulders, aiding and abetting a tendency toward natural cracks, root action, freezing, and thawing. Many rocks get a reddish-brown surface in a short amount of time. First they turn color, and then they fall apart.

So the air might attack a fresh surface from the outside, while water flowing through cracks can alter the inside. Water is a great solvent—it tries to dissolve everything it meets. Then toss in the sandblasting effect of fierce winds carrying grit, plus roots breaking down cracks, spring floods, and extreme baking heat, and you can see that rocks at the surface will not last long. It can be hard to tell the difference between an old sandstone and a much younger basalt if they both have weathered to the same rusty brown.

Where to Go

Most of the United States contains interesting collecting opportunities, and there is a rockhounding title in the FalconGuide Rockhounding series for all the important states or regions. Use this book as a companion title to help identify what you find in the field.

Many field guides exist to help you locate material; use this identification guide as a companion to the specific field guide for your state.

About This Guide

This guide was organized with beginners in mind. The most common rocks are listed first, with a table to help you narrow down what you're looking for and pictures if you are still baffled. The thinking here was that if you don't know the name, organizing the contents alphabetically won't help much.

Because the United States is rich in fossils, many quite valuable or scientifically interesting, there is a section for various fossils. Remember that you cannot remove vertebrate fossils from public lands without a paleontological permit, but leaves, petrified wood, and insects are all there for the collecting.

For the minerals, there are sections for common minerals and common metals important for mining. The common minerals form most of our rocks and are rarely important to miners, except as indicators and to determine structures. The minerals in the metals section are generally valuable, come from mining districts, and are usually either attractive or economically interesting. Most of the highly scientific jargon and measurements stayed out; the information you need, like streak, hardness, and color, stayed in.

This exotic "7-Upite" would never make it into the collecting bag of some rockhounds. Your mileage may vary.

Finally, there are gemstones. These are the most valuable or highly sought rocks and minerals from across the country. You can try to get them on your own, using one of the appropriate Falcon-Guides as a companion to this title, or you can haunt the shows and shops and online retailers. What you collect, and the order you collect in, is up to you.

Remember that for all the "leaverite" you encounter—rocks that someone suggests you should "leave 'er right" there—you never know what might appeal to you. Even glass can be interesting.

Hardness

A lot of the time, especially when you're starting out, you can't always tell what you have. There are several ways to key out a rock or crystal, such as the streak, hardness, luster, color, and weight.

The hardness test is one of the most elementary ways to sort out samples. Try to keep a few of these common specimens on

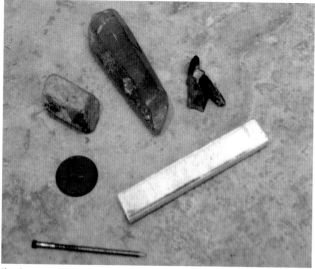

If you keep a scratch plate, a nail, a penny, some feldspar, a quartz crystal, and some topaz handy, you should be able to identify most minerals.

hand, and if possible memorize the ten standard listings. I've included the "nonstandard" testing materials in italics.

Mohs #	Material
1	Talc
2	Gypsum
2½	*Fingernail*
3	Calcite
3½	*Copper penny*
4	Fluorite
5	Apatite
5½	*Knife blade, nail, or window glass*
6	Orthoclase feldspar
6½	*Pyrite; steel file*
7	Quartz; agate, chert, chalcedony, and jasper; streak plate
7½	*Garnet*
8	Topaz
9	Corundum
10	Diamond

Be a Safe and Considerate Collector

Before going any farther, let's talk about safety and etiquette in the field. The biggest safety tip is to never push your car beyond its limits, and bring appropriate tools in case you need to do a simple roadside repair. Upgrade your tires, and be religious about vehicle maintenance. Your expeditions can take you to the end of some very long and dangerous roads. Walking out is no fun.

Don't treat private property like public lands; get all the permissions you need beforehand or move on. Obey all signs, control your kids and dogs, pick up litter, and take only what you need when collecting. Consider bringing along an extra bucket to carry out some of the trash you may encounter when searching road cuts. Bring a shovel and use it. Bury toilet paper deep. Monitor any smokers in the group, and make sure they use a "butt jar" to collect their cigarette butts rather than flicking them out car windows and across the landscape.

Artifacts

Under the Antiquities Act, archaeological resources are protected from removal on public lands. You need to be aware of the possibilities, particularly when you pick up agate, jasper, flint, chert, chalcedony, and obsidian samples in the field. There is some controversy over collecting arrowheads. The 1979 Archaeological Resources Protection Act stated in section 3, subsection (g): "Nothing in subsection (d) of this section shall be deemed applicable to any person with respect to the removal of arrowheads located on the surface of the ground." This is sometimes called the "Carter Clause," as President Jimmy Carter was himself an arrowhead collector, and at the time there was even a Boy Scouts merit badge for collecting arrowheads from the surface of the ground.

Yet this is still a gray area. Digging, screening, excavating, and other harmful activities are clearly illegal, somewhat immoral, and always scientifically reckless. Casual surface collecting would seem to be a different story, but it's not. Subsequent regulations and other legal interpretations appeared to move arrowheads

The chances are good that at least one of these fragments was once a tool used by Native Americans.

back into protected status. If you find an intact arrowhead, your safest bet is to photograph it and leave it. Barring that, you could photograph it in situ, record the global positioning system (GPS) coordinates, carefully wrap it in a protective plastic bag, and then report it to local authorities. They'll tell you if it is significant enough to arouse their interest.

The next challenge is that while you're certainly going to know an intact arrowhead when you see it, many other artifacts are hard to identify. In fact, fragments and chips are quite likely to wind up in your collecting bag, and you might not even know that you have a scraper, hand ax, or similar tool until you get it home and clean it up. In that case, you again owe it to yourself to find out. Contact the nearest Bureau of Land Management (BLM) or USDA Forest Service office or a local natural history museum and let them know what you found.

Additional Reference Materials

This book can help you with about 99 percent of the geology you'll see in the field, because most of the visible rocks at

Earth's surface are common sedimentary rocks. Igneous rocks are also extremely plentiful, and once you know the basic rules for those, you'll be able to screen them out as you look for nicer specimens.

The National Audubon Society Field Guide to North American Rocks and Minerals is a huge treasure trove of 200+ color photographs. It's glossy, fits in a big pocket, and probably takes a beating in the field. I leave mine at home.

For web-based research, I recommend the Minerals Database (mindat.org). It's a noncommercial online database of almost 39,000 mineral names, more than 300,000 photos, and 125,000 localities; and it is community editable. It has some GPS coordinates and can always use more.

The US Geological Survey (USGS) operates the Mineral Resources Data System (MRDS) (http://mrdata.usgs.gov/mrds/), with excellent information on professional papers and publications related to mining districts, fossil locales, and more. It has interactive maps and plenty of data. The USGS map site (usgs.gov/pubprod/) has topo maps for the United States.

I highly recommend the Gator Girl website at gatorgirlrocks.com, with resources for every level of rockhound. Another excellent online resource is Wikipedia, which has entries for many rocks and minerals. I also strongly recommend thoughtco.com's geology page (thoughtco.com/geology-4133564) for a variety of general geologic information, including maps. Also, geology.com is a good site to skim around for info.

I'm a huge fan of the Roadside Geology series because they give even the most amateur geologist a general idea about the rock exposures passing by your windshield. I also recommend that every rockhound become familiar with geology maps, learn to use a GPS device, take great pictures, and share what you find via your favorite social networking site.

State Designations

It helps to know what fossils, rocks, minerals, and gems are recognized as "official" for a state. Some regions have recognized

Land Use

Some of the most lasting collectibles from your trips into the field come home via a camera rather than a rock bag. The vistas you'll encounter in national parks are normally composed of common rocks in dramatic formations, and they are worth every penny you spend to get in. Some of my favorite national parks involve geology in some important way.

National Parks & Monuments

Grand Canyon National Park (Arizona)—a slice through time, exposing dramatic rock formations

Yellowstone National Park (Wyoming)—active geysers, dramatic vistas, and lots of volcanic rocks

Carlsbad Caverns National Park (New Mexico)—limestone caves with impressive features

Yosemite National Park (California)—impressive granite and diorite cliffs

Petrified Forest National Park (Arizona)—unbelievable gem-quality tree trunks lying on the ground

Mammoth Cave National Park (Kentucky)—impressive subterranean experience

Big Bend National Park (Texas)—mines, battlefields, and stunning views near the Rio Grande

Great Smoky Mountains National Park (Tennessee)—stunning views and mountain vistas

Devil's Tower National Monument (Wyoming)—impressive volcano "throat"

Fossil Butte National Monument (Wyoming)—amazing fossils from the Green River Formation

Agate Fossil Beds National Monument (Nebraska)—great mammal fossil collection

Dinosaur National Monument (Utah)—a giant cliff of Mesozoic fossils left in place

Florissant Fossil Beds National Monument (Colorado)—
noted for fossilized insects and plants
John Day Fossil Beds National Monument (Oregon)—
prized for mammals, nuts, and leaves

Leave your geology pick and collecting bag in your car when you visit these protected lands. Here are the public land policies specific to national parks:

- No collecting of any kind is allowed. No pebbles as souvenirs, no yard rocks, no samples.
- No camping except in designated areas.
- No hiking except on official trails, sometimes requiring proper paperwork, such as backcountry permits.
- No driving except on designated roads.

Public lands exist for all citizens to enjoy. The USDA Forest Service and the BLM are the primary land stewards in the United States; other agencies, such as state, county, and wildlife agencies, also control minor lands. Obey their rules and enjoy the scenery! For more information about the National Park Service, visit nps.gov.

unique material, found only in their state. Others have claimed material that is common but prominently recognized within their borders. Sometimes there might be an interesting story behind how a fossil or mineral earned special status. Wikipedia lists all the state gems, fossils, rocks, and minerals at http://en.wikipedia .org/wiki/List_of_U.S._state_minerals,_rocks,_stones_and_gem stones; it's a handy reference when you're planning your trips.

PART I
ROCKS
Igneous Rocks: Extrusive

Use the matrix below as a starting point for understanding how extrusive igneous rocks are organized.

Name	Color	Grain Size	Composition	Useful Characteristics
Matrix of Extrusive Igneous Rocks				
Andesite	Medium	Fine or mixed	Medium silica lava	Plagioclase feldspar with pyroxenes
Basalt	Dark	Fine or mixed	Low silica lava	Has no quartz in rock, but can contain seams of agate
Dacite	Light	Mixed	Medium silica	Plagioclase with hornblende
Felsite*	Light	Fine or mixed	Medium to high silica	Rich in quartz and feldspar
Obsidian	Dark	Fine	High silica	Glassy
Pumice	Light	Fine	Frozen lava froth	Small bubbles
Rhyolite	Light to medium	Fine or mixed	Very high silica	Quartz is common
Scoria	Dark	Fine	Runny lava froth	Large bubbles
Tuff	Light	Mixed	Andesite, rhyolite, or basalt	Ash fragments; sometimes fused by heat. Can contain petrified wood or fossils.

*Not covered in this guide

Andesite

Common andesite is not as dark as basalt and usually has no flow lines.
PHOTO COURTESY OF THE RICE NORTHWEST MUSEUM OF ROCKS AND MINERALS

Group: Igneous; extrusive
Mineralogy: Intermediate; feldspars, pyroxene, and/or hornblende
Key test(s): Fine grained with pyroxene and plagioclase
Likely locale(s): Volcanic mountain ranges

Andesite is a common form of lava, usually lighter in color than basalt. It dominates many volcanic mountain ranges, as its higher silica content causes it to pile up better as opposed to flood basalts that flow for miles. This rock is named for the Andes Mountains, where it is a key component of those volcanoes. Andesite often exhibits a platy structure, appearing in outcrop as though stacked, but that isn't a telltale clue. Andesite is the extrusive equivalent of diorite, matching that rock in chemical composition. This lava is typically fine grained, with small crystals of plagioclase feldspar (such as andesine), hornblende, pyroxene (such as augite or diopside), and biotite. It makes a fine decorative rock, weathering from its light gray appearance when fresh to a dark gray or black color.

Andesite occurs in volcanic terrain throughout the western United States. New England and the Midwest contain less-significant exposures.

Basalt

Ropy basalt flows at Craters of the Moon National Monument near Arco, Idaho.

Group: Igneous; extrusive

Mineralogy: Plagioclase feldspar, olivine, and pyroxene

Key test(s): Fine grained, often with microscopic crystals only; can form dramatic columns

Likely locale(s): Volcanic regions

Basalt occurs in many parts of the Rocky Mountain states. It appears gray or light black when fresh but quickly weathers to a tan, yellow, or brown surface. It can be difficult to identify because it tends to be fine grained, so look for clues in the deposit itself. Vesicles are common; these holes sometimes fill with agate, chalcedony, or opal in small, round "blebs" that can weather out and are collectible. Basalt can also host zeolites, quartz veins, and calcite veins under the right conditions. Andesite and rhyolite are thick, viscous lavas that build up faster, while basalt is runny and can flow hundreds of miles. While still referred to as extrusive, basalt can form in large sills and cool slowly enough to create impressive polygons, referred to as columnar jointing. These colonnades are collectible as impressive towers.

Basalt is common in the western United States, especially across southern Idaho's Snake River plain, such as at Craters of the Moon National Monument. The Columbia River basalts of the Pacific Northwest rival some of the biggest basalt formations in the world. California, Texas, Minnesota, the New England states, and many other states host basalt flows.

Dacite

Dacite comes in a variety of forms, often appearing as dikes cutting through basalt or as domes in volcanic craters.

Group: Igneous; extrusive

Mineralogy: Plagioclase feldspar, quartz, biotite, hornblende, and augite

Key test(s): Hornblende and plagioclase in fine-grained matrix

Likely locale(s): Volcanic ranges; domes, dikes, or sills

In composition, dacite ranks between andesite and rhyolite, typically featuring much larger crystals and a coarse, patchy appearance. Dacite is not common, so it can be difficult to become familiar with it. The color is quite variable, ranging from white, gray, and rarely black to pale red or brown; dacite may even occur as deeper reds and browns. The matrix frequently appears oxidized. Mineralogically, dacite contains the same components as granodiorite, with plagioclase feldspar, quartz, biotite, hornblende, and augite. Dacite usually occurs as ignimbrite or tuff, or as domes, dikes, and flows; it can display some flow banding, and can resemble obsidian.

Cascade volcanoes in Oregon and Washington contain significant dacite. Nevada, California, and Arizona also contain notable deposits.

Obsidian

Obsidian is common in many western states.

Group: Igneous; extrusive
Mineralogy: Rhyolitic
Key test(s): Glassy luster, sharp edges, conchoidal fracture pattern
Likely locale(s): In rhyolite flows

Obsidian has no crystal structure—it chilled too fast for the atoms to align in a crystal lattice—so technically it is a glass and thus a rock, not a mineral. It's an easy rock to identify because it has a glassy luster when freshly chipped. When weathered or rolled in streams and rivers, however, with no fresh surface, obsidian can resemble every other rhyolite pebble on the outside. Other than the classic shiny black luster, look for evidence of flow, such as banding. The familiar fracture pattern is "conchoidal," or shell-like; knappers have long used obsidian's tendency to break predictably to fashion amazing stone tools. The most common obsidian is jet black, but there are other colors and forms.

Throughout the western Basin and Range Province, rockhounds use the term *Apache tears* for fragments of obsidian formed in perlite and varnished by desert winds. Oregon, California, Nevada, and Wyoming are all known for interesting obsidian deposits.

Pumice

Pumice is so frothy and full of air that it floats.
PHOTO COURTESY OF THE RICE NORTHWEST MUSEUM OF ROCKS AND MINERALS

Group: Igneous; extrusive
Mineralogy: Andesitic to rhyolitic
Key test(s): Floats; soft
Likely locale(s): Volcanic mountain ranges

Look for pumice in known volcanic regions, especially any area with recent events. Pumice is light enough to float, thanks to its frothy, glassy texture, which traps air and makes each rock buoyant. Pumice has a unique texture and feel that makes it easy to identify, and it can show dramatic flow lines. Many violent pyroclastic events spew pumice with ash and other material, causing pumice fragments to show up in breccias, tuffs, and other ash deposits. Pumice rafts are also common after volcanic eruptions in a marine environment. Collectors don't need more than a sample or two for a complete collection unless considering industrial uses for pumice, including yard rock, potting soil, and abrasives.

Pumice is common in younger volcanic regions in the western United States, such as in Oregon, Washington, Idaho, and California. Fresh pumice is rare, as it is fragile; it is more likely to wind up in sandstone, siltstone, and mudstone.

Rhyolite

Yellowstone Canyon slices through a dramatic sequence of rhyolite flows.

Group: Igneous; extrusive

Mineralogy: Quartz, feldspar, biotite, hornblende, with accessory magnetite

Key test(s): Quartz crystals in fine-grained matrix

Likely locale(s): Common as pyroclastic flows and caldera fill

Rhyolite is another extremely common igneous rock, so learning to distinguish it from andesite and basalt is worthwhile. Its high silica content usually means rhyolite is lighter in color. In general, rhyolite is light gray, but it can appear yellow, pale yellow, and from pale red to deeper red. Rhyolite can also form in welded, compacted bands with dramatic color varieties, sometimes called wonderstone, which is carvable and takes a polish. Rhyolite cools faster than basalt and usually appears to be composed of fine-grained but distinguishable crystals. It is sometimes glassy. Flow banding is common, and crystals often show alignment under a hand lens, as do any small vesicles present.

The most famed rhyolite locale is Yellowstone National Park, a resurgent caldera that represents one of the biggest volcanic dangers on Earth, capable of erupting 200 cubic miles of material. Most western states feature rhyolite deposits; the New England states also contain significant deposits.

Scoria

Scoria is heavier and more solid than pumice but has a similar texture.
PHOTO COURTESY OF THE RICE NORTHWEST MUSEUM OF ROCKS AND MINERALS

Group: Igneous; extrusive

Mineralogy: Usually similar to basalt

Key test(s): Rusty, like cinders; numerous vesicles; heavier than pumice

Likely locale(s): Volcanic regions

Sometimes called cinder, scoria resembles pumice because both typically contain numerous air pockets, or vesicles. While pumice floats, however, scoria sinks in water. Where pumice is soft and easy to crush, scoria is usually harder and more rigid. The name comes from the Greek word for rust, and the orange-red rusty color is a common characteristic of some forms of scoria. Thus, the name *scoria* refers to a rock with a certain color or texture rather than a specific mineralogy. The tops of lava flows, where considerable froth sometimes takes place, can harden into scoria. Similarly, some frothy ejected material from eruptions can harden as scoria. Scoria is often used in landscaping and in barbecue grills.

Scoria is common in volcanic regions such as Idaho's Snake River plains. Most of the southwestern states and Basin and Range areas contain scoria.

Tuff

Most of the "painted hills" in the western deserts are tuff deposits.

Group: Igneous; extrusive
Mineralogy: Basalt to rhyolite
Key test(s): Soft, unless welded; angular clasts
Likely locale(s): Downwind from volcanic ranges

Tuff, or hardened (lithified) volcanic ash, is a common fragmental deposit around and downwind from volcanoes. Also called tephra, the term can be a bit of a catchall for the pyroclastic material ejected and deposited downwind. Tuff can be somewhat loose and unconsolidated, or it can be welded into harder varieties, all the way to a type called "picture jasper," which is lined with swirls and patterns. Tuff is usually tan to dark brown in color but can occur as pink, yellow, green, or even purple. There are varieties of tuff known as crystal tuff, containing mineral crystals; lithic tuff, with rock fragments; and welded tuff, in which pumice fragments were hot enough to compress in glassy wisps. Geologists also denote rhyolitic, andesitic, basaltic, ultramafic, and trachyte. Tuff is usually found as a breccia, with a fine- to medium-grained matrix, and can contain large angular material such as pumice, volcanic bombs, and other debris. Tuff beds can be prime fossil-hunting locales containing petrified wood, leaves, bones, and other organic debris trapped by violent volcanic events.

Many of the "painted hills" in the western United States are tuffs and volcanic ash. Tuff deposits are common throughout the western states, as well as New England, Virginia, Texas, and Michigan.

Igneous Rocks: Intrusive

Below is a table comparing the various intrusive igneous rocks.

Matrix of Intrusive Igneous Rocks				
Name	Color	Grain Size	Composition	Features
Diorite	Medium to dark	Coarse	Low calcium	Limited quartz
Dunite*	Green	Coarse	Olivine dominant	Dense; 90+% olivine
Gabbro	Medium to dark	Coarse	High calcium	No quartz; limited olivine
Granite	Light, often pink	Coarse	Feldspar, quartz, mica, and hornblende	Wide range of color; coarse grain size
Kimberlite*	Dark	Coarse	Source of diamonds	Found in "pipes"
Pegmatite	Any	Very coarse	Usually granitic	Dikes; small intrusions
Peridotite	Dark; often greenish	Coarse	Olivine present	Dense; 40+% olivine
Porphyry	Any	Mixed	Large grains of feldspar, quartz, olivine, pyroxene	Large grains in a fine-grained matrix
Pyroxenite*	Dark	Coarse	Pyroxene	Rich in pyroxene
Syenite*	Light	Coarse	Mostly feldspar and minor mica	Like granite but no quartz
Tonalite*	Light to medium; salt-and-pepper	Coarse	Plagioclase feldspar and quartz plus dark minerals	Limited alkali feldspar

*Not covered in this guide

Diorite

Granodiorite from the northern Rockies
PHOTO COURTESY OF THE RICE NORTHWEST MUSEUM OF ROCKS AND MINERALS

Group: Igneous; intrusive
Mineralogy: Feldspar, hornblende, biotite, and pyroxene
Key test(s): Dark smears, unlike granite; often in dikes and sills
Likely locale(s): Mountainous terrain

Diorite is a medium- to coarse-grained intrusive igneous rock, made up of common minerals such as plagioclase feldspar and pyroxene. It is usually medium grained, so crystals are recognizable under a hand lens, but it can be coarse, with zoning common. Diorite is usually gray to dark gray but can be lighter, and a green or brown tint isn't out of the question. Salt-and-pepper colors are very common. Compared to granite and granodiorite, plain diorite is not common; it often occurs as dikes, sills, and stocks at the margins of large granite batholiths. Because diorite is relatively hard, it takes a good polish. Ancient artisans used diorite to carve inscriptions—the Code of Hammurabi is carved in a black diorite; the Rosetta Stone is carved in a granodiorite.

Just about every major area that contains granite also contains diorite. The New England states, the Rocky Mountain region, Arizona, California, and Washington are all good places to look.

Gabbro

Gabbro typically has a very coarse texture.
PHOTO COURTESY OF THE RICE NORTHWEST MUSEUM OF ROCKS AND MINERALS

Group: Igneous; intrusive
Mineralogy: Feldspar, hornblende, biotite, and magnetite
Key test(s): Dark with coarse grains
Likely locale(s): Mountainous terrain

Gabbro is an interesting intrusive rock, distinguished by a very coarse texture and a range of dark colors. One helpful characteristic is the presence of green olivine, which contrasts with the usual black or dark gray color. Gabbros can be dark red, however, so the color test isn't precise. Gabbro sometimes makes up the base unit of massive, layered intrusions, such as the famed Stillwater Complex in Montana. Perhaps the most famous layered intrusive with gabbro at the base is the storied Bushveld Complex of South Africa, the source of much of the world's platinum. These complex intrusions are often rich in chrome as well. Layering is sometimes visible in fresh exposures, but gabbro tends to weather faster, which can also aid in field identification. Other names for gabbro include diabase, dolerite, and black granite. Gabbro takes a polish and can be fashioned into dark countertops, monuments, and statues.

The gabbros of Montana around the Stillwater Complex are the most noteworthy, but gabbro occurs in many western states. Washington, Oregon, Idaho, and California all feature significant gabbros; New England and eastern states, as well as Michigan and Minnesota, are also important gabbro locales.

Granite

Granite can be pink, gray, or a mix called salt-and-pepper.

Group: Igneous; intrusive
Mineralogy: Quartz, feldspar, hornblende, biotite, and magnetite
Key test(s): Often pinkish; exfoliation
Likely locale(s): Mountainous terrain

Granite is one of the most common and recognizable intrusive igneous rocks. It is usually coarse grained and is primarily composed of feldspar, hornblende, and quartz. It can be massive or display zoning, depending on how close it came to the exterior of the intrusion. It is sometimes a salt-and-pepper rock; some granites have significant white feldspar, while others have a lot more dark hornblende. Granite may occur as minor intrusions, as larger plutons, or as huge batholiths taking up tens of thousands of square miles. One field clue is the tendency of granite to weather by exfoliation, where thin sheets peel off like the skin of an onion. Granite isn't highly collectible, but artisans cut granite into tabletops and counters or use it in buildings and monuments because granite takes a high polish. Probably the most interesting aspects of granite are its association with economic ore deposits, often found along the margins of granite intrusions, and the presence of pegmatites.

Granite mountains and batholiths are common across the United States—from Maine to Georgia; across the Midwest, such as at Mount Rushmore; across the West; and all the way to Alaska.

Pegmatite

Pegmatite features large crystals of feldspar, mica, schorl, and other minerals.
PHOTO COURTESY OF THE RICE NORTHWEST MUSEUM OF ROCKS AND MINERALS

Group: Igneous; intrusive

Mineralogy: Quartz, feldspars, altered olivine, rare garnets, orthopyroxene, and chrome diopside

Key test(s): Rare garnets; very coarse crystals

Likely locale(s): Older continental crust

Pegmatites sometimes host the most prized of gems, such as rubies, emeralds, and sapphires, so their potential reward makes them important to learn about. The problem is that they get complicated quickly. The word *pegmatite* refers to its coarse texture with large crystals, sometimes resembling a patchwork quilt. Pegmatites are usually found at the margins of large granite bodies and contain the same minerals, such as alkali feldspar and quartz. Other minerals include quartz, tourmaline, topaz, mica, apatite, lepidolite, and monazite. There are varieties, such as granite pegmatite and nepheline syenite pegmatite, but pegmatites are all usually white due to the amount of feldspar present. Pegmatites can also appear light yellow, tan, or even gray. Zoning is common, with vugs and cavities highly sought for large crystals of quartz, smoky quartz, topaz, and other prizes.

Famous pegmatite belts and outcrops occur in many states, including New England and along the East Coast. California, Montana, Colorado, Idaho, and Arizona are noteworthy in the western states.

Peridotite

Chunky peridotite with visible olivine from New Mexico
PHOTO COURTESY OF THE RICE NORTHWEST MUSEUM OF ROCKS AND MINERALS

Group: Igneous; intrusive
Mineralogy: Olivine and pyroxene
Key test(s): Green; coarse
Likely locale(s): Ancient terrains

Peridotite is a dark, dense rock made up of olivine and little else. Peridot is the gem variety of olivine and lends its name to this rock. Peridotite can be as much as 90 percent olivine; higher concentrations would make it a dunite. The olivine is usually present as large crystals; thus, a coarse texture is another key distinguishing characteristic. It can form in layers or as massive structures. Peridotite is one of the most exotic rocks available in the field, as it is closely associated with Earth's interior and is probably the primary rock in the upper mantle. Since olivine reacts quickly with water and oxygen, peridotite is not very stable; upon exposure to the atmosphere, it quickly starts converting into serpentinite.

Peridotite locales include Montana's Stillwater Complex and Wyoming's State Line kimberlite district. California's Trinity Mountains, Klamath Mountains, and the Sierras are all significant. Oregon, Massachusetts, and Minnesota also host interesting peridotite locales.

Porphyry

Andesite porphyry showing scattered large, dark crystals
PHOTO COURTESY OF THE RICE NORTHWEST MUSEUM OF ROCKS AND MINERALS

Group: Igneous; intrusive
Mineralogy: Typically basalt or rhyolite
Key test(s): Contains large quartz or feldspar crystals
Likely locale(s): Volcanic regions

Rockhounds call a rock a porphyry when it appears to contain notable large crystals in a fine-grained matrix. The chemical composition of the large crystals and the matrix may be similar, but something caused two different phases of cooling in these rocks. The large crystals are usually 2 millimeters or greater in size and are typically distinguishable as either quartz or feldspar. This apparently represents a time when the liquid magma cooled slowly, deep inside the earth, because large crystals typically take a long time to form. Subsequent events led to the still-liquid magma rising rapidly to the surface and then cooling quickly so that the large crystals formed. Most igneous rocks seen in the field show this texture if studied closely, but striking porphyries will catch your eye.

This texture occurs in many volcanic rocks throughout most volcanic regions, especially in Arizona and Texas. New England states such as Massachusetts and New Hampshire also contain notable porphyries.

Metamorphic Rocks

Use the matrix below as a starting point for understanding how metamorphic rocks are organized. Foliation (from the Latin *folium*, meaning "leaf") refers to the sheet-like or lined structure common in schists and gneisses.

Name	Hardness	Foliation	Grain Size	Color	Other
Amphibolite	Hard	Foliated	Coarse	Dark	Hornblende
Argillite	Hard	Relicts	Fine	Mixed	Common
Eclogite*	Hard	Nonfoliated	Coarse	Red and green	Dense; garnet and pyroxene
Gneiss	Hard	Foliated	Coarse	Mixed	Banded
Greenstone	Medium	Relicts	Fine	Light	Common
Hornfels	Hard	Nonfoliated	Fine or coarse	Dark	Dull, opaque
Marble	Medium	Nonfoliated	Coarse	Light	Calcite or dolomite by acid test
Migmatite*	Hard	Foliated	Coarse	Mixed	Melted
Mylonite*	Hard	Foliated	Coarse	Mixed	Crushed and deformed
Phyllite	Soft	Foliated	Fine	Dark	Shiny and crinkly
Quartzite	Hard	Nonfoliated	Coarse	Light	Quartz—no fizzing
Schist	Hard	Foliated	Coarse	Mixed	Large, deformed crystals
Serpentinite	Soft	Nonfoliated	Fine	Green	Shiny, mottled
Slate	Soft	Foliated	Fine	Dark	Striking sound
Soapstone*	Very soft	Foliated	Fine	Light	Greasy

Not covered in this guide

Amphibolite

Amphibolite is composed mostly of hornblende and is thus very dark and heavy.

Group: Metamorphic; regional
Mineralogy: Amphiboles such as hornblende and actinolite; feldspar
Key test(s): Crystalline, often salt-and-pepper
Likely locale(s): Metamorphic terrain

Amphibolite is a general term for dark, dense metamorphic rocks that contain mostly members of the amphibole family. Geologists refer to an "amphibolite facies" to denote a particular pressure and temperature range required to produce the minerals inside. There may be bands or lines present, and it can be easy to confuse amphibolite with a banded gneiss. There is little or no quartz present in amphibolite. There are at least a dozen different amphiboles, but generally hornblende and actinolite are the two major minerals present. There are many different rock types that can metamorphose into amphibolite, including basalt and sedimentary rock. Note that garnets and rubies are often associated with amphibolite.

There are numerous amphibolite locales in the older rocks of New England and the East Coast. Wisconsin, Idaho, California, and Alabama also offer amphibolite locales.

Argillite

Massive argillite boulder from Missoula, Montana
PHOTO COURTESY OF THE RICE NORTHWEST MUSEUM OF ROCKS AND MINERALS

Group: Metamorphic; regional
Mineralogy: Muscovite mica, chlorite; pyrite a frequent accessory
Key test(s): Foliation; visible crystals; density
Likely locale(s): Metamorphic terrain

Argillite is a common term for any fine-grained sedimentary rocks, such as siltstone, that have begun to metamorphose. Think of argillite as a mud or clay that has hardened into a rock. Argillites can be highly variable in composition but generally have high concentrations of aluminum and silica. Argillites sometimes mix with shale, making the distinction difficult. Argillite tends to be massive, but bands and zones of different composition give them an almost sedimentary look. Related rocks include black jadeite, catlinite, pipestone, dickite, and haidite.

One notable locale for argillite is the Superbelt area of northeast Washington, northern Idaho, and neighboring Montana, where it is common. California, New England, Arizona, and Minnesota also contain interesting argillite exposures.

Gneiss

Banded gneiss from Georgetown, Colorado
PHOTO COURTESY OF THE RICE NORTHWEST MUSEUM OF ROCKS AND MINERALS

Group: Metamorphic; regional
Mineralogy: Muscovite mica, chlorite; pyrite a frequent accessory
Key test(s): Foliation; visible crystals; density
Likely locale(s): Metamorphic terrain

Increasing metamorphism ⟶

Mudstone	Shale	Slate	Phyllite	Schist	Gneiss

Gneiss (pronounced nice) is another extremely common meta-morphic rock. In metamorphic terrain, it can make up most of the pebbles and cobbles of a river or creek. Gneiss is medium to coarse grained, and the minerals align tightly. This rock shows off light and dark banding in an infinite variety, but it is usually harder than schist and thus sometimes easily distinguished from those rocks. Note, however, that there is no acknowledged cutoff between schist and gneiss. Banding and layering are key characteristics; rocks and outcrops demonstrate folding on both large and small scales. Augen gneiss contains rolled "eyes," or rounded masses. Hardness is highly variable: Scratch tests show a range of 6 to 7 for hard, quartz/feldspar layers, down to 2½ to 3 for biotite-rich layers. Key components include orthoclase feldspar, quartz, biotite, and hornblende. Accessories include almandine garnet, corundum, and staurolite. Look for intense folding, light-and-dark banding, and hard, erosion-resistant rocks at the heart of mountain ranges.

Gneiss is common in old metamorphic terrains across the United States—New England and other eastern states feature significant exposures. California, Idaho, and Wisconsin also host considerable gneiss.

Greenstone

Greenstone is a sign of regional metamorphism.

Group: Metamorphic; regional
Mineralogy: Muscovite mica, chlorite; pyrite a frequent accessory
Key test(s): Foliation; visible crystals; density
Likely locale(s): Metamorphic terrain

Greenstone is a dense, fine-grained rock that probably started life as a gabbro or, more likely, a seafloor lava flow such as pillow basalt or plain basalt. It has undergone regional metamorphism to cook into a plain, green appearance. It is generally massive, with occasional evidence of its basalt parentage, such as vesicles. It has a dull luster and is green, pale green, or occasionally yellow green. The hardness for greenstone is about 5 to 6, but it can vary depending on which minerals dominate the specimen. Prospectors quickly learned to locate greenstone outcrops, as the telltale green color was easy to spot. Its presence signifies regional metamorphism, and finding it usually leads to discovery of quartz and calcite veins that could transport economic ore deposits. However, it can be very common in some areas and thus is not a sure indicator of economic ore deposits.

Greenstones were important in Wyoming, Montana, and Colorado gold-bearing regions. New England, Washington, and California also feature significant greenstone belts.

Hornfels

Spectacular lime hornfels with large biotite blades

Group: Metamorphic; regional
Mineralogy: Muscovite mica, chlorite; pyrite a frequent accessory
Key test(s): Foliation; visible crystals; density
Likely locale(s): Metamorphic terrain

Hornfels is a common metamorphic rock in some terrains, but it can be difficult to identify. The name refers to the erosion-resistant Matterhorn in the Swiss Alps, where the rock forms prominent peaks. Biotite hornfels is usually black or at best dark gray, while lime hornfels is lighter in color. Both varieties are very dense. Hornfels can be very fine grained and massive, with very few signs of relict bedding. Hardness ranges from 6 to 7, similar to gneiss. When pounded with a hammer, hornfels breaks into angular splinters, but that isn't a surefire field test.

The best places to look for hornfels are around the margins of big batholiths and intrusions, such as the Kaniksu Batholith and the Idaho Batholith, where there was plenty of heat. California's Triassic marine rocks around Death Valley, the Mojave Desert, and Mono Lakes also feature hornfels rocks, and there are scattered occurrences throughout the United States.

Marble

Brilliant white Yule marble from Colorado highlights several Washington, DC, monuments.

Group: Metamorphic; regional
Mineralogy: Muscovite mica, chlorite; pyrite a frequent accessory
Key test(s): Foliation; visible crystals; density
Likely locale(s): Metamorphic terrain

There are dozens of varieties of marble, but in general it is simply a limestone or dolomite that has been highly metamorphosed, usually into a fine- to medium-grained specimen with a soft, vitreous luster. Marble is normally white, sometimes brilliantly so, but it is often darker, thanks to black, green, red, pink, or even yellow staining from various impurities. It often contains veins of darker material that originated in the limestone bed as clay, sand, silt, or chert. This produces a lined, or "marbled," look. Because limestone deposits are often quite large, it's logical that marble is usually massive, with foliation, banding, and streaking all common. Marble has a hardness of only about 3 but can range from 2 to 5. Calcite marble effervesces with cold hydrochloric acid. Chemists must first pulverize dolomitic marble and then heat the acid to get the expected fizzing action.

Marble occurs throughout the United States. Significant quarries and exposures occur in the New England states; up and down the eastern seaboard; and in Alabama, California, and Idaho.

Phyllite

Phyllite outcrop showing thin, platy structure and crumbly texture

Group: Metamorphic; regional
Mineralogy: Muscovite mica, chlorite; pyrite a frequent accessory
Key test(s): Foliation; visible crystals; density
Likely locale(s): Metamorphic terrain

Increasing metamorphism ⟶

Mudstone	Shale	Slate	Phyllite	Schist	Gneiss

Phyllite has undergone more heat and pressure than slate but has not yet hardened into a schist, so it tends to be crumbly and easy to crush by hand. It tends to be medium to dark gray but can turn black if enough carbon was present in the original sedimentary rock. It is mostly mica, such as muscovite and chlorite, and the abundance of mica is what contributes to the foliation. Geologists term this kind of rock "friable" because it is so crumbly. Phyllite grades into schist, and the boundaries can be subtle.

Phyllite is common in the Appalachians of eastern North America and throughout the eastern seaboard. Colorado, Idaho, and California also host significant deposits, as it is a common rock.

Quartzite

Typical quartzite pebble, fracturing across bedding planes

Group: Metamorphic; regional
Mineralogy: Muscovite mica, chlorite; pyrite a frequent accessory
Key test(s): Foliation; visible crystals; density
Likely locale(s): Metamorphic terrain

Quartzite is a common pebble in many streams and rivers because it is hard and strong. Quartzite has a vitreous luster and often occurs as a round, white rock, especially if pure. Other forms of quartzite can be light gray, and it can even run to pink or light brown. Quartzite is usually fine grained, but sometimes crystals are actually visible and the rock can be medium grained. Quartzite deposits are typically thick, massive units, resistant to erosion and forming prominent bluffs. Quartzite will break across bedding planes, not along them. The hardness is typical of quartz, at 7, but it can be softer if significant amounts of feldspar and calcite are present. This rock takes a nice polish and has some industrial applications, including landscape rock, tiles, flooring, and even ballast. Some collectors mistake quartzite pebbles for agate because they share similarities but reserve the term *agate* for translucent, banded quartz.

Quartzite pebbles and outcrops occur throughout New England and the eastern seaboard states. There are numerous significant exposures and formations throughout the western states, the Southeast, and the upper Midwest. Quartzite, Arizona, is surrounded by good exposures.

Schist

Schist with almandine garnets and muscovite mica from Emerald Creek, Idaho

Group: Metamorphic; regional
Mineralogy: Muscovite, biotite, kyanite, and glaucophane varieties
Key test(s): Foliated mica; garnets; density
Likely locale(s): Metamorphic terrain

Increasing metamorphism ⟶

Mudstone	Shale	Slate	Phyllite	Schist	Gneiss

Schist is one of the more common metamorphic rocks, and it is relatively simple to identify. Outcrops often contain striking folds that look like wavy lines. Schist is usually medium to coarse grained, with mica being obvious. Gray and brown are common colors, but schist can run to white and yellow, depending on how much iron is present to stain the rock. The hardness of a typical schist is highly variable. It can range to 1 on the Mohs scale if considerable talc is present and to around 5 with more hornblende. If quartz is dominant, the rock's hardness can reach 7 or so. Chlorite and albite schists have undergone low-grade heat and pressure, while the presence of garnet and epidote represent medium-grade metamorphism. These are usually referred to as greenschists. Further heat and pressure result in kyanite—a blue, bladed mineral—and then staurolite, noted for its twinning. These blueschists are prized for their hardness in jetties.

Schist is common throughout the metamorphic zones of the United States, such as in New England and along the East Coast. The Appalachians, Rockies, Sierra Nevada, and many other mountain chains contain abundant schist.

Serpentinite

Serpentinite is a jumble of several related green serpentine minerals.

Group: Metamorphic; regional
Mineralogy: Muscovite mica, chlorite; pyrite a frequent accessory
Key test(s): Green, greasy, streaky, and soft
Likely locale(s): Metamorphic terrain

Serpentinite is a collective term for about twenty different magnesium iron phyllosilicates commonly found in varying percentages. Serpentine is the most well-known constituent, but other common minerals include antigorite, lizardite, and chrysotile. Many of these minerals are amphibole asbestos, so use caution. This rock is usually green, with streaks of black, yellow, white, or gray. It is very dense, which distinguishes it from soapstone. The best field test is serpentinite's slick, greasy feel, which is a key characteristic. The structure is usually massive, if jumbled, and serpentinite is famous for "slickensides," where microfaulting and zones of movement result in polished, "slick" surfaces. The hardness for serpentinite is usually 4 or less on the Mohs scale, sometimes as low as 2½, depending on how much hard quartz is present. Serpentinite usually occurs in small pods, lenses, and layers, with steep tilting common. It is carvable, and its varied hues make for interesting sculptures, even if it is harder than soapstone.

Serpentine is the state rock for California, and Serpentine Canyon dominates the upper Feather River. Oregon, Washington, North Carolina, and Massachusetts are some of the many states that contain serpentinite occurrences.

Slate

Dark slate with small, brassy pyrite crystals
PHOTO COURTESY OF THE RICE NORTHWEST MUSEUM OF ROCKS AND MINERALS

Group: Metamorphic; regional
Mineralogy: Muscovite mica, chlorite; pyrite a frequent accessory
Key test(s): Foliation; visible crystals; density
Likely locale(s): Metamorphic terrain

Degree of metamorphism ⟶

Mudstone	Shale	Slate	Phyllite	Schist	Gneiss

Slate started life as a sedimentary rock, most likely a mudstone, but it experienced heat and pressure that caused the minerals to reorient themselves. At this level of metamorphism, the principal minerals are micas, such as muscovite and chlorite. Pyrite crystals and cubes, sometimes weathered to limonite, are common in some slates. Slate is sometimes confused with phyllite, but if you have visible mica flakes and the rock is crumbly and easy to break, it's a phyllite. Slate is usually dark gray, thanks to the presence of graphite from carbonaceous parent material. Slate makes great flagstones and building material and still shows up in high-end billiard tables.

Many US regions have undergone enough metamorphism to host slate deposits, but commercial quarries in the Slate Valley of New York and Vermont produce very high-quality material. Experts also cite the Monson slate of Maine, and many eastern states host excellent shale deposits. In the West, only California is noted for top material. Washington, Alabama, Michigan, and Tennessee also contain significant slate formations.

Sedimentary Rocks

Use the matrix below as a starting point for understanding how sedimentary rocks are organized.

Name	Hardness	Grain Size	Composition	Features
Arkose*	Hard	Coarse	Quartz and feldspar	Quartzy sandstone
Breccia	Hard or soft	Very coarse	Mixed rocks and sediment	Sharp-edged conglomerate
Chert	Hard	Fine	Chalcedony	No acid fizz
Claystone	Soft	Very fine	Clay-rich hardened mud	Hard-to-see particles
Coal	Soft	Fine	Carbon	Black; burns
Concretions	Hard, round	Fine	Lime-rich matrix; organic material inside	Round "cannonballs"
Conglomerate	Hard or soft	Very coarse	Mixed rocks and sediment	Rounded rocks glued together
Coquina	Soft	Coarse	Fossil shells	Bits and pieces
Dolomite	Soft	Coarse or fine	50% dolomite crystals	Fizzes if powdered
Limestone	Soft	Fine	Calcite	Fizzes
Phosphorite	Soft	Coarse or fine	Phosphorus	Gray or white
Sandstone	Hard	Coarse	Clean quartz	Grainy
Shale	Soft	Fine	Clay minerals	Splits in layers
Siltstone	Hard	Fine	Very fine sand; no clay	Gritty
Wacke/ greywacke	Hard or soft	Mixed	Mixed sediments with rock grains and clay	Gray or dark and dirty

Not covered in detail in this guide

Breccia

Breccia is composed of angular clasts that have not had their corners rounded off.

Group: Sedimentary
Mineralogy: Large clasts in cement matrix
Key test(s): Rounded pebbles
Likely locale(s): Non-marine

Unlike conglomerates, breccias consist of broken rock pieces that are still angular and sharp edged. This is usually evidence that the source material is located nearby. These rocks are similar to tuff and volcanic ash and are usually lighter in color. They contain mixed material, but the general rule is that the rock fragments are coarse, angular, and varied. Breccias often occur in thick, massive beds, possibly interbedded with fine-grained tuff. Breccias are sometimes evidence of explosive forces, such as meteorite-impact breccias, and from volcanic explosions. Breccias can also depict evidence of significant movement along faults.

Volcanic and sedimentary breccias are common across the United States.

Chert

Red chert is easy to confuse with jasper, so look for evidence of bedding to verify chert.

Group: Sedimentary
Mineralogy: Silica
Key test(s): Hardness: 7; organic nature; ooze origins
Likely locale(s): Nodules in limestone

Like most quartz incarnations, chert has a hardness of 7 on the Mohs scale. It is commonly white in color—when it is usually referred to as flint—but banded chert can take on reds, yellows, and even greens. Chert is a by-product of organic ooze from the seafloor hardening into a rock, while jasper results from circulating silica solutions, usually in basalt. At high magnification, some cherts display tiny skeletons. Flint is a common form of chert occurring as black or dull-gray chunks and nodules, especially as nodules in limestone deposits. Chert is dense, and smooth when polished, but it can feel very rough where exposed. Bedded cherts can be big enough to slice with a rock saw and polish into cabochons.

Many states contain chert deposits; it is common among limestone and dolomite layers. Alabama, New York, New Jersey, and Pennsylvania are the notable states for chert along the East Coast. Oklahoma boasts good chert deposits; Nevada, California, and Idaho are leaders among the western states.

Claystone

Claystone is a more scientific name for mudstone. These formations often contain fossil shells in marine environments.

Group: Sedimentary; clastic
Mineralogy: Very fine
Key test(s): Visible bedding planes
Likely locale(s): Marine

Claystone is another common sedimentary rock, with very fine grains but lacking the ability to split easily. If it split better, it would be a shale, but claystone is massive and doesn't show bedding planes. Modern classifications refer to shale and argillites as mudstones, as they are a mix of silt and clay. Sedimentologists refer to grain sizes less than 0.002 millimeter as clay, and claystones contain at least 50 percent of such particles. These rocks vary in color, such as light gray and light brown when fresh, but they can be darker, depending on the parent rock that eroded. Under a microscope you can look for clay, such as kaolinite, feldspars, and quartz grains, plus mica flakes, but at a very small scale. Fossils tend to be soft in these rocks and often require an immediate coating of polyvinyl acetate to preserve.

Claystone is common across the United States, grading into sandstone, siltstone, and conglomerates, sometimes in alternating bands.

Coal

Common coal has a dark, dull to shiny appearance; the lower the grade, the messier it is.

Group: Sedimentary
Mineralogy: Carbon
Key test(s): Oily smell; dusty; light heft
Likely locale(s): Sedimentary and metamorphic terrain

Increasing metamorphism ———→

Peat	Lignite	Bituminous	Anthracite	Graphite	Carbon

Coal actually spans the boundary between a sedimentary rock and a metamorphic rock. Even at high grades, coal is very soft; its overall hardness ranges from 1 to 2½. Once coal bakes into the anthracite stage, it is less greasy, but coal dust is always a problem due to coal's tendency to be brittle and friable. Further metamorphism leads to graphite and pure carbon, but diamonds exist in kimberlite, not from metamorphosed coal. The higher the quality, the fewer relicts are present, and color can get more interesting, with purple iridescence. High-grade coal is up to 98 percent hydrocarbons, but sulfur and nitrogen are usually present.

Coal is notable in Pennsylvania, Ohio, Kentucky, West Virginia, and Tennessee. Many other states contain deposits, with Wyoming a leading producer. Utah, Texas, New Mexico, and several other western states host interesting coal-bearing formations.

Concretions

Fossil crab prepared from a lime-rich concretion

Group: Sedimentary
Mineralogy: Clastic environment usually
Key test(s): Round, heavy
Likely locale(s): Sandstone

Like geodes, concretions are a "lottery ticket" of the geology world. (**Note:** The terms *nodule* and *concretion* are often used interchangeably.) The outside appearance may be dull and drab, but the inside may contain marvelous treasures. Concretions are typically round or oval structures that can range from tiny, pea-size orbs to large, beach ball–size or larger monsters. A concretion typically forms when some organic debris starts rolling around in a lime-rich mud within an active marine bay or lagoon. The mud sticks to the organic material and soon forms a small, round ball. Inside could be a fossil or, in some cases, a septarian nodule with minerals such as barite. Or iron oxide could replace the calcium carbonate, creating an iron oxide nodule. Some parts of the sea-floor are covered with rounded manganese-iron concretions.

The Pierre Shale, Frontier Formation, and Bearpaw Formation stretch across much of Montana, Wyoming, and Colorado. The Glendive, Montana, area hosts ammonites and also septarian nodules—complex concretions that have formed, dried, cracked, and refilled, sometimes with barite. Oregon and Washington feature crab-bearing concretions, and many recent marine deposits across the United States contain fossil-bearing concretions.

Conglomerate

Conglomerate contains rounded pebbles, often cemented in a very strong matrix.
PHOTO COURTESY OF THE RICE NORTHWEST MUSEUM OF ROCKS AND MINERALS

Group: Sedimentary
Mineralogy: Large clasts in cement matrix
Key test(s): Rounded pebbles
Likely locale(s): Non-marine

Conglomerates typically occur as hard, haphazardly sorted beds of large, rounded pebbles glued together in a fine- to medium-grained matrix. If the pebbles are still angular and edgy, geologists call the rock a breccia. This is a clastic rock, and the clasts themselves can be big or small. By definition, the rocks and pebbles are greater than 2 millimeters in diameter, but they can range to large cobbles and boulders. Because conglomerates typically form in non-marine waters that are very active, the clasts are usually hard material such as quartzite or chert, which can survive that much churning. The matrix of a conglomerate is often silica or calcite, sometimes making this rock extremely tough and resistant to erosion. Fossils are rare in conglomerates because the high velocity required to move these rocks tends to destroy anything soft or fragile.

Conglomerates are very common across the United States, as the bottom layers in a typical sandstone often contain the largest clasts. Since most states contain sedimentary rocks, conglomerates are scattered across the continent.

Coquina

This Florida beach is well on its way to becoming a future coquina limestone.

Group: Sedimentary
Mineralogy: Chemical deposition
Key test(s): Fizz test; plentiful fossils
Likely locale(s): High-energy marine zones

Coquina is a loose term for a form of limestone composed chiefly of fossil fragments. Sometimes nicknamed "shell hash," it is easy to spot. It contains varying sizes and types of fossil shells, but the agreed-upon rule is that the average size of the fragments should be over 2 millimeters. Coquina is sometimes soft when first removed, but it hardens over time. The shell fragments often show sorting, bedding, and even orientation. A lime-rich cement may or may not do a good job of holding the grains together.

Coquina limestone is used for building material in Florida.

Dolomite

Common dolomite from Wyoming's Bighorn Formation

Group: Sedimentary
Mineralogy: Chemical
Key test(s): Fizz test
Likely locale(s): Common

Dolomite, sometimes called dolostone to differentiate it from crystalline dolomite, is typically tan or light gray, but varieties can range all the way to pink and dark gray. It is typically dense like limestone, its cousin, but there is usually less evidence of grains or fossils. Typically, dolomite has a very fine texture because it doesn't have the shell fragments or oolites of limestone. Limestone and dolomite share many characteristics, but the big difference is that dolomite has substituted more magnesium for calcium. It can appear as bands and grade slowly into limestone. The best test for separating dolomite from limestone is with hydrochloric acid: Dolomite will only fizz if the acid is hot and the material has been ground into a powder, while limestone is much more reactive. Dolomite ends up as fertilizers high in calcium and magnesium.

Dolomite is common in southeast Idaho, Wyoming, and Colorado, especially adjacent to limestone rocks. The southeastern states, especially Alabama, contain significant dolomite, and those zones stretch to Missouri, Tennessee, and Pennsylvania. Texas and Oklahoma contain interesting dolomite, as do many western states.

Limestone

Madison limestone from Wyoming shows bedding planes.

Group: Sedimentary
Mineralogy: Chemical deposition
Key test(s): Fizz test; fossils
Likely locale(s): Old sea basins

Limestone is one of the most common sedimentary rocks, and it comes in many varieties. Chalk is fine grained, derived from the skeletons of tiny sea creatures, while coquina contains large, abundant shell fragments. Travertine is typically banded and colorful, while oolitic limestone refers to tiny orbs, or oolites, which are very small concretions. The term *marl* describes a limestone with a high percentage of silicates. Limestone is often marked in the field by pitted and pockmarked outcrops. The hardness for limestone is in the range of 3 to 4. The most reliable test is fizzing in cold, diluted hydrochloric acid. Typically, limestones are light gray; but if iron is present, they can stain reddish, trending to darker gray.

Limestone is common throughout the United States. Florida, Georgia, Alabama, and Mississippi all have significant formations; so do the New England and Midwest states. Even western states with significant igneous rocks have interesting limestone deposits.

Phosphorite

Phosphate-rich phosphorite from the Phosphoria Formation of southeast Idaho and western Wyoming

Group: Sedimentary
Mineralogy: Chemical
Key test(s): Fizz test
Likely locale(s): Sedimentary basins

Phosphorite is typically tan or dark white. It is characteristically dense like dolomite (also called dolostone), but there is usually less evidence of grains or fossils. Phosphorite has a very fine texture because it doesn't have the shell fragments or rounded oolites of limestone. Phosphorite and dolomite share many characteristics, but the big difference is that dolomite has substituted more magnesium for calcium, while phosphorite is rich in the element phosphorous. It can appear as bands and grade slowly into shale or dolomite. The best test for separating dolomite from phosphorite is with hydrochloric acid: Dolomite will fizz if the acid is hot and the material has been ground into a powder.

Phosphorite from the Phosphoria Formation in Utah, southeast Idaho, and western Wyoming supplies most of the phosphate fertilizer in the United States. Montana and Texas contain sporadic phosphorite deposits.

Sandstone

Red sandstone cliffs at Mesa Verde, Colorado

Group: Sedimentary; clastic
Mineralogy: Large quartz or feldspar clasts in cement matrix
Key test(s): Rounded pebbles
Likely locale(s): Non-marine

Sandstone is one of the most common sedimentary rocks, occurring throughout the United States. Sandstones in the field frequently are interbedded with claystone, limestone, shale, and other sedimentary rocks. By definition, all sandstones are marked by a grain size of 0.05 to 2.0 millimeters, which makes it medium grained compared to mudstone and siltstone but not coarse like a conglomerate. Sandstone displays great variety in color, occurring in hues of gray, brown, red, yellow, and even white. What holds the clasts or grains together in a sandstone is usually silica, but calcite or even iron oxide will also serve. Sandstone is relatively easy to shape and fashion into buildings and walls, and it is often used as a decorative stone.

Just about every US state contains some sandstone exposures. Sandstones from the Cretaceous era are common in the Rocky Mountain states, and locally they contain abundant dinosaur fossils. Two of the more famous formations are the Morrison and Hell Creek Formations.

Shale

Shale with weak bedding planes from the Gros Ventre Formation in Wyoming

Group: Sedimentary
Mineralogy: Muscovite mica, chlorite; pyrite a frequent accessory
Key test(s): Foliation; visible crystals; density
Likely locale(s): Sedimentary basins

Increasing metamorphism ⟶

Mudstone	Shale	Slate	Phyllite	Schist	Gneiss

Shale is a sedimentary rock, but it has started to harden in distinguishable ways. Varieties include oil shale, calcareous shale, carbonaceous shale, and others. Shale is usually dark but not always, as its color depends on the source rock from which it hardened. Shale is noted for the way it fractures into plates, along bedding planes, where evidence of past water action such as waves, ripples, cracks, and footprints are all observable. Shale beds are easy to spot, although they can easily be confused with platy andesite. Look for evidence of fossils or water features for one differentiator. Note that some rockhounds use the slang "shale" for *any* unwanted rock, no matter what its true composition is.

Shale is common throughout the United States wherever there are significant layers of sedimentary rocks.

Siltstone

Siltstone with fossil leaves, from the Chuckanut Formation of northwest Washington

Group: Sedimentary; clastic
Mineralogy: Fine–very fine
Key test(s): Visible bedding planes
Likely locale(s): Marine

Think of siltstone as a very fine-grained sandstone. It is a common sedimentary rock that ranges from light gray and light brown in color, but it can be darker if more organic material or iron staining is present. It is usually dense, and the particles are at least 50 percent silt-size, or 0.002 to 0.0063 millimeter. Clay is anything less than 0.002 millimeter. Old-school geologists could tell the difference in grain size by chewing a sample carefully with their teeth. Clay and shale with no silt don't have a gritty feel, while siltstone, mudstone, and coarse shale do feel gritty. Look at your sample under a microscope to see if it contains mostly small particles of silt or clay. Bedding planes are often easy to see, especially if wet or marked by fossils. Look for clay, such as kaolinite, feldspars, and quartz grains, plus mica flakes, but at a very small scale. Fossils tend to be soft in these rocks and often require an immediate coating of polyvinyl acetate to preserve.

Siltstone is common across the United States, grading into sandstone, claystone, and coarser rock units.

Wacke/Greywacke

Common greywacke showing poor sorting and reddish iron staining

Group: Sedimentary
Mineralogy: Clastic
Key test(s): Clay oxidation
Likely locale(s): Sedimentary basins

Greywacke, a specific type of wacke (rhymes with whacky), is a sedimentary rock composed of poorly sorted, rounded, or broken bits of common rock in a fine clay matrix. Easy to mistake for an odd breccia, greywackes are usually gray, hence the name, but can be dark gray or reddish with the presence of iron. There is typically no sorting in a greywacke, giving it an unusual appearance among breccias, conglomerates, and sandstones. Quartz and feldspar grains are common, but the grain size and structure of a greywacke varies quite a bit. Think of it as a dirty sandstone, which typically dates to Paleozoic rocks.

Exposures in the United States are restricted to folded mountain ranges that don't contain a lot of limestone, such as in Humboldt, Washoe, and Pershing Counties in Nevada. Another locale is at Baker Beach near the Presidio in California.

Fossil Algae (stromatolites)

Sliced and polished stromatolites from the Green River Formation near Wamsutter, Wyoming
PHOTO COURTESY OF THE RICE NORTHWEST MUSEUM OF ROCKS AND MINERALS

Group: Invertebrate

Mineralogy: Usually replaced by quartz; sometimes iron

Key test(s): Visible wavy lines

Likely locale(s): Commonly marine

Stromatolites are fossilized algae mats, representing the evolutionary advance from single one-celled organisms to colonies. They are among the oldest forms of life on Earth, dating to the Precambrian era, yet they still exist today. They can occur as nodules and round blobs or as wavy lines, and colors range from tan to gray and even purple. Some forms erode quickly, but hard pieces will take a polish.

Stromatolites are common across the Belt Supergroup from Idaho's panhandle region. Midwest states such as Michigan, Wisconsin, and Minnesota host fossil stromatolites, sometimes associated with banded iron deposits. Wyoming also boasts collectible specimens from the South Pass area and younger Eocene specimens in the Green River Formation. Alaska and Quebec also feature these ancient fossils.

Fossil Coral

Petoskey stones from Michigan are fossils of Hexagonaria percarinata, *a common coral from the Devonian.*

Group: Invertebrate
Mineralogy: Hard body parts replaced by calcite
Key test(s): Plantlike appearance
Likely locale(s): Marine

This unique rugose coral hails from near Petoskey, Michigan. While unremarkable in its raw state, the fossilized coral is actually hard enough to take a good polish. The individual cells in the colony are usually six-sided, but not always. Glacial action apparently plucked these rocks from their source in the Gravel Point Formation; dunes and beaches near Lake Michigan still yield material.

Petoskey stones are the state fossil of Michigan. Other fossil coral locales include Texas and much of the Southeast, including Florida.

Fossil Fish

Mass mortality plate of a fossil fish, Knightia alta, *from the Green River Formation in Wyoming*
PHOTO COURTESY OF THE RICE NORTHWEST MUSEUM OF ROCKS AND MINERALS

Group: Vertebrate

Mineralogy: Usually replaced by quartz

Key test(s): Bone structure

Likely locale(s): Marine and freshwater sedimentary rocks

The fossil bones of fish are protected by law, as are all vertebrates, and only licensed paleontologists connected to museums can collect such fossils from public land. Private land is a different story, and there are many fee-dig operations across the Rocky Mountain states. The fossil fish *Knightia* was the equivalent of today's herring and sardines and was an abundant food source in much of the Green River Formation. It is Wyoming's state fossil.

Excellent fee-based fossil quarries surround Fossil Butte National Monument near Kemmerer, Wyoming. These quarries yield Eocene-age fish from thinly bedded limestone. Florissant Fossil Beds National Monument is another well-regarded locale. New York, Illinois, Iowa, Michigan, and other eastern states contain older fish fossils predating the "age of reptiles."

Fossil Insects

Fossil beetle, Cybister explanatus, *from Rancho La Brea Tar Pits in Los Angeles, California*
PHOTO COURTESY OF THE RICE NORTHWEST MUSEUM OF ROCKS AND MINERALS

Group: Invertebrate and vertebrate
Mineralogy: Insect hard parts usually replaced by carbon
Key test(s): Visible insect parts, such as wings and legs
Likely locale(s): Delicate thin-bedded sedimentary rocks

Fossil insects are rare, requiring delicate conditions and gentle ash, silt, or mud deposits. The tar pits at La Brea have yielded excellent fossil insects. Worldwide, most well-preserved insects come from amber, but there are only limited amber deposits in the United States. So fossil insect hunters must search sedimentary rocks, particularly from ancient lakes, and concentrate on fine shales and siltstones. Unlike petrified wood, where quartz is required, fossil insect exoskeletons are composed of chitin and calcium carbonate, and they convert to carbon or charcoal.

Fossil Bowl is a reasonable fee-dig site at Clarkia, Idaho, and contains occasional insect fossils from the Miocene era. Fossil Butte National Monument near Kemmerer, Wyoming, is famous for the variety of its Eocene-age insect population. Florissant Fossil Beds National Monument in Colorado contains abundant Eocene insect fossils—more than 1,500 separate species to date. Florissant sponsors digs during the summer.

Fossil Leaves

Fossil fern, Polypodiaceae, *from the Pennsylvanian period, found in the Llewellyn Formation of Schuylkill County, Pennsylvania*
PHOTO COURTESY OF THE RICE NORTHWEST MUSEUM OF ROCKS AND MINERALS

Group: Invertebrate
Mineralogy: Carbon remains
Key test(s): Appearance
Likely locale(s): Freshwater or marine rocks

Unlike for collecting vertebrate fossils, there are very few restrictions on collecting fossil leaves. Collectors can specialize in particular types, periods, ecosystems, and more. In addition, some fossil leaf locales, or quarries, are incredibly abundant, with great variety. Many of the siltstone, sandstone, and mudstone formations of the United States feature fossil leaf material.

One term to become familiar with is *Lagerstätte*, German for "storage place." Such sites yield specimens that still contain DNA and boast their original color when first exposed to oxygen.

Fossil Mammals

Fossil saber-toothed cat, Smilodon, skull from the La Brea Tar Pits of Los Angeles, California

Group: Vertebrate

Mineralogy: Usually replaced by quartz

Key test(s): Bone structure

Likely locale(s): Marine and freshwater sedimentary rocks; traps and caves

When alive, mammal bone contains the mineral apatite and collagen tissue; upon death the collagen decays. When the bone is fossilized, silica replaces the collagen or occupies the voids left behind, making the bone hard and strong. As with all vertebrate fossil collecting, you should mark a spot you suspect contains mammal fossils and show it to a nearby museum curator or university paleontologist.

Sedimentary rocks throughout the United States contain numerous mammal fossils. John Day Fossil Beds National Monument in Oregon is a good example of Cenozoic mammals; the Hagerman Fossil Beds in Idaho offer more recent specimens. New England and the Great Lakes states contain numerous examples of what scientists call Pleistocene megafauna, which seem to have died off around the same time human hunters expanded their skills.

Fossil Reptiles

Tyrannosaurus rex *skull from the Wyoming Dinosaur Center in Thermopolis*

Group: Vertebrate

Mineralogy: Usually replaced by quartz

Key test(s): Bone structure

Likely locale(s): Marine and freshwater sedimentary rocks

The first dinosaur fossil excavated in the United States came from north-central Montana in 1855. In the years that followed, a kind of "bone wars" broke out between leading paleontologists as they competed to find and name as many species as possible. The Mesozoic sedimentary rocks of the Rocky Mountain states yielded thousands of specimens. Fee-dig opportunities still exist for aspiring dino-diggers. Hell Creek, Montana, is just one example (paleo trek.com).

Numerous sedimentary formations containing dinosaur fossils dot the Rockies and the Dakotas. By far the best book on the subject for beginners is *Cruisin' the Fossil Freeway,* wonderfully written by Dr. Kirk Johnson, former chief curator at the Denver Museum of Nature and Science, and whimsically illustrated by noted artist Ray Troll. Their wall map of the western United States is a must-have for aspiring paleontologists. Most major natural history museums boast at least one good fossil reptile.

Fossil Sea Creatures

Fossil crinoids, Eretmocrinus granuliferus, *from Humboldt County, Iowa*
PHOTO COURTESY OF THE RICE NORTHWEST MUSEUM OF ROCKS AND MINERALS

Group: Invertebrate
Mineralogy: Hard body parts replaced by calcite
Key test(s): Plantlike appearance
Likely locale(s): Marine

Not all ancient sea creatures in the fossil record are common shells such as clams and snails. Although they are known as sea lilies, crinoids (pronounced KRĪ-noid, with a long *i*) are not plants but animals, the food-gathering arms attached to a strong, hollow stalk. Crinoids are one of the most dominant Paleozoic fossils. The cylindrical stalks often break laterally, forming round or even five-sided sections that were once known as "Indian beads" because they were naturally easy to string. These fossils date to the Ordovician period and are sometimes the major component of limestone beds.

The state fossil of Missouri, crinoids are primarily found throughout the Midwest and Pennsylvania, as well as Texas. Many other states yield crinoid fossils.

Fossil Shells

Fossil sea urchins, Eupatagus floridanus, *from the Ocala limestone of Florida*
PHOTO COURTESY OF THE RICE NORTHWEST MUSEUM OF ROCKS AND MINERALS

Group: Invertebrate
Mineralogy: Shells replaced by calcite
Key test(s): Calcite
Likely locale(s): Marine

Clams and snails are the most common shell material recorded in the fossil record, but there are many other interesting creatures preserved in marine rocks. Crustaceans such as crabs, crayfish, and lobsters may show up inside lime-rich concretions or as carbon-rich imprints in sandstone. Shark vertebrae are common in some marine rocks along the Pacific coast. Limestones, mudstones, and sandstones are the most common rocks to host fossil shells, but there are fascinating variations. Some Midwest quarries, such as those near Indianapolis, Indiana, contain fossil clams (pelecypods) where the calcium carbonate has been replaced by pyrite. Some fossil molds of ammonites and snails (gastropods) are known to have been replaced with agate, or even carnelian agate.

Fossil shells show up in sedimentary rocks across North America, especially—but not exclusively—near ocean shores.

Fossil Teeth

Fossil shark tooth, Carcharodon megalodon, *from South Carolina*

Group: Vertebrate
Mineralogy: Tooth enamel and apatite replaced by quartz
Key test(s): Hardness, appearance
Likely locale(s): Marine

Shark teeth are often the only things that remain in the fossil record. Shark teeth are reasonably abundant because sharks regularly shed their teeth as new ones grow, and scientists have described about 3,000 species. When alive, shark teeth are mostly the mineral apatite plus collagen; if fossilized, both undergo replacement with phosphates. Coloring usually derives from minerals in the host sandstone.

Fossil teeth, especially from the ancient shark *Carcharodon megalodon*, occur in marine sedimentary rocks in various places across the United States. Famed locales in Florida include along the coast near Tampa Bay and in the muddy bottom of the Peace River. The Cooper River in North Carolina is another celebrated locale. Shark Tooth Mountain in California is now predominately private land. Elsewhere, shark teeth can occur in concretions or scattered through marine sediments.

Fossil Wood (petrified wood)

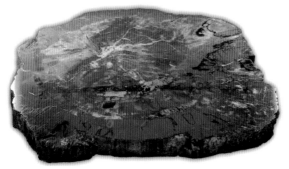

Slabbed and polished petrified wood from the Petrified Forest, Arizona
PHOTO COURTESY OF THE RICE NORTHWEST MUSEUM OF ROCKS AND MINERALS

Group: Plants
Mineralogy: Usually replaced by quartz
Key test(s): Visible wood structure, especially growth rings
Likely locale(s): Primarily sedimentary; also ash deposits

Petrified wood is the common name for mineralized wood material. The mineralization is some form of silica, or quartz, including agate, chalcedony, opal, or jasper. Some specimens could contain two or more forms of quartz, so the term *petrified* is still useful, as it is a generic term for "turned to stone." Since this term is so generic, it could also account for opalized wood, which is often frosty white, pink, or blue and reflects replacement by almost pure chalcedony. Being somewhat brittle, most forms of petrified wood disintegrate faster in rocky streambeds, so most recovered samples are geologically younger.

Petrified wood is quite common across the western United States, wherever there are ash beds accompanying lava flows. The Blue Forest near Farson, Wyoming, is a famed locale frequented by rockhounds.

Trace Fossils

Fossil dinosaur tracks from Tyrannosauripus coelurosauria *found in Jurassic sediments*

Group: Any

Mineralogy: Varies; no physical remains of organism

Key test(s): Evidence of life

Likely locale(s): Sedimentary rocks

Trace fossils are a huge challenge for rockhounds. In the case of the fossilized dinosaur tracks shown here, you need to imagine a muddy area that was trampled by a large reptile and then possibly filled in with slightly different sediments. The mud had to be hard enough to hold the animal's print, soft enough to deform. The only clue is the distinctive, three-toed cast, which is easy to miss unless you're looking for it. There is no bone, no skin, or any other information—it's all guesswork. Other examples of trace fossils include burrows in sediments or wood left by wormlike creatures.

PART III
MINERALS

Common Minerals

Actinolite

Actinolite has distinctive, delicate green crystals.

$Ca_2(Mg,Fe)_5Si_8O_{22}$
Family: Amphibole silicate
Mohs: 5–6
Specific gravity: ~3
Key test(s): Green, fibrous, brittle
Likely locale(s): Metamorphic terrain

Actinolite is usually green, ranging from pale green to dark green, sometimes with yellow tinting. It can display as long, matted blades, usually quite brittle, and the blades can be parallel or create spectacular sprays. Like so many other minerals, actinolite occurs in a series, in this case with one end being rich in magnesium and the other rich in iron, and near-infinite percentages between the two extremes. Actinolite sits in the middle; magnesium-rich actinolite is tremolite, while iron-rich actinolite gets the unimaginative name ferro-actinolite. These rocks are a source of asbestos, so beware. Related rocks include nephrite jade.

Actinolite is common across the western United States, including Washington, California, and Arizona. Montana, South Dakota, and Wisconsin also feature locales. New England, North Carolina, Virginia, and Tennessee all contain occurrences of actinolite.

Agate—Carnelian

Banded carnelian agate with drusy quartz at the center

Quartz, SiO$_2$
Family: Cryptocrystalline quartz
Mohs: 6½–7
Specific gravity: 2.65
Key test(s): Banding; translucence; orange-red color
Likely locale(s): Veins and blebs in basalt

Agate is a term loosely used by rockhounds to describe any clear or translucent quartz, but the term is interchangeable with *chalcedony*, which is the preferred term among mineralogists. However, agate hunting, agate picking, and agate digging are such a part of our lexicon that the term will never go away. Agate is one of the most popular and collectible forms of quartz, and it comes in a wide variety, including lace agate, plume agate, moss agate, and banded agate. Carnelian agate is highly prized, and the more-vivid orange or red forms are instantly recognizable. If the piece is so red as to be opaque, it is called sard.

Carnelian agate in multiple forms occurs across the United States but is especially noteworthy in basalt terrains, where it forms as seams and fillings.

Agate—Dendritic

Translucent agate with dendritic inclusions from Crane, Montana, on the Yellowstone River
PHOTO COURTESY OF THE RICE NORTHWEST MUSEUM OF ROCKS AND MINERALS

Quartz, SiO$_2$
Family: Cryptocrystalline quartz
Mohs: 6½–7
Specific gravity: 2.65
Key test(s): Banding; translucence
Likely locale(s): Veins and blebs in basalt

Rockhounds use the term *dendritic agate*, or "moss" agate (some-times called "crazy lace agate"), especially when the dark den-drites are thick and banded. Dendrites are usually black, formed from manganese oxide, and often display interesting patterns and inclusions. Often rockhounds won't know if there are den-dritic patterns until slicing and polishing their material. The "rind," or outside, of such agates tends to be plain as well, so finding this kind of agate is difficult.

Southwestern Montana contains notable deposits of dendritic agate. Wyoming's Platte County also contains excellent dendritic agate. Other famed agate locales in the Rocky Mountain states include fluorescent Sweetwater agates in Fremont County, Wyo-ming. The famed agate beds at Graveyard Point, reached via Marsing, Idaho, yield moss agate. Grand and Larimer Counties in Colorado have yielded moss agate in the past. Colorado's Weld County contains agate similar to that found in the famed beds at Fairburn, South Dakota.

Agate—"Dryhead"

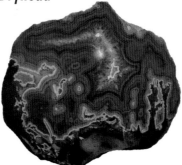

Dryhead agate from Montana's Yellowstone River
PHOTO COURTESY OF THE RICE NORTHWEST MUSEUM OF ROCKS AND MINERALS

Quartz, SiO$_2$
Family: Cryptocrystalline quartz
Mohs: 6½–7
Specific gravity: 2.65
Key test(s): Banding; translucence
Likely locale(s): Veins and blebs in basalt

Unlike common agate and chalcedony, dryhead agate (sometimes called "crazy lace agate") contains interesting banding patterns that may reflect repeated saturation by silica solutions as the material formed. Inconsistent chemical impurities such as iron give the bands distinctive colors and result in outstanding patterns when cut and polished.

Montana's Yellowstone River still contains notable pieces of dryhead agate.

Amethyst

Plate of purple amethyst crystals showing deep color and sharp terminations
PHOTO COURTESY OF THE RICE NORTHWEST MUSEUM OF ROCKS AND MINERALS

Quartz, SiO_2
Family: Crystalline quartz
Mohs: 7
Specific gravity: 2.65
Key test(s): Purple; harder than fluorite
Likely locale(s): Quartz veins

Amethyst is a form of crystalline quartz, like smoky quartz, only amethyst is typically purple, sometimes vividly violet. Iron impurities impart the purple color. Like common quartz, amethyst reaches a hardness of 7 on the Mohs scale. Crystals are hexagonal and often sharply terminated. Note that the distinctive purple color can fade under intense sunlight.

The West Coast states of Washington, Oregon, and California all host amethyst deposits, either as crystal-lined geodes or as scepters and points. Nevada and Colorado contain noteworthy deposits; Montana, Arizona, and Wyoming also contain amethyst. Missouri, Arkansas, Ohio, North Carolina, and New England also boast interesting amethyst locales.

Apatite

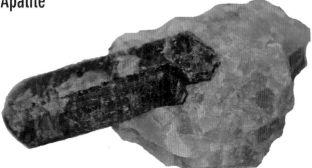

Well-formed green apatite crystal
PHOTO COURTESY OF THE RICE NORTHWEST MUSEUM OF ROCKS AND MINERALS

Calcium phosphate, $Ca_5(PO_4)_3(F,Cl,OH)$
Family: Phosphates
Mohs: 5
Specific gravity: 3.1–3.2
Key test(s): Hardness test
Likely locale(s): Pegmatites

Apatite is the name given to a group of phosphate minerals such as fluorapatite and chlorapatite, so the chemical formula can vary according to how much of one element or another is present. As expected, the color varies. Darker green, purple, or violet is common, as are red, yellow, and pink. Apatite is a common mineral in igneous rocks; larger crystals occur in pegmatites. Fluorite has similar coloration but isn't as hard, and the crystal habit is cubic; quartz is harder. Interestingly, apatite is the hardening agent for bones and teeth.

Apatite is scattered across North America, with notable occurrences in most of the western states. Arkansas and Wisconsin host apatite locales, as do eastern states from Florida to Maine.

Aragonite

Aragonite needles are another form of calcium carbonate.
PHOTO COURTESY OF THE RICE NORTHWEST MUSEUM OF ROCKS AND MINERALS

Calcium carbonate, CaCO$_3$

Family: Carbonates

Mohs: 3½–4

Specific gravity: 2.93

Key test(s): Fibrous appearance

Likely locale(s): Fossil shells, stalactites

Aragonite is another form of calcium carbonate, but unlike its related cousin calcite, aragonite usually occurs thanks to biological processes. Crystals are often white or milky white, but with impurities a vast range of colors are possible. Some mollusks form their shells completely with aragonite; others alternate between calcite and aragonite. Aragonite is not strictly biological, however. For example, the stalactites of Carlsbad Caverns are aragonite. The flowering pattern is often distinguishing. Aragonite is not very stable in a geologic sense—there are no deposits of aragonite known that are older than the Carboniferous period, which ended about 300 million years ago.

Aragonite is common around limestone deposits of the United States. There are notable occurrences in the western states, especially California and Colorado. Texas and Florida are two southern states with aragonite locales. The New England states and Pennsylvania also host interesting deposits.

Augite

Large augite crystals are rare, but augite is a common component of igneous rocks.

Silicate, $Ca,Na(Mg,Fe,Al)(Al,Si)_2O_6$

Family: Silicates

Mohs: 5–6

Specific gravity: 3.2–3.6

Key test(s): Black, stubby crystals

Likely locale(s): Volcanic dikes

Augite is a common mineral in the pyroxene group, related to wollastonite, diopside, hedenbergite, pigeonite, and others. Augite typically forms short, stubby crystals in the monoclinic system, but they are usually too small to detect without a hand lens. Large crystals do form occasionally and can be striking and quite shiny, but they pit quickly when exposed to air. Augite has a greenish-white streak, which is a good field test.

Augite typically occurs in igneous rocks, sometimes as an essential mineral in dikes. It is common in the volcanic rocks of the western states—especially Oregon—but also occurs in Alabama, Kentucky, and New England.

Barite

Rare blue barite crystals from Stoneham, Colorado
PHOTO COURTESY OF THE RICE NORTHWEST MUSEUM OF ROCKS AND MINERALS

Barium sulfate, BaSO$_4$
Family: Sulfates
Mohs: 3–3½
Specific gravity: 4.3–4.6
Key test(s): Heavy; fluorescence
Likely locale(s): Layered sedimentary rocks; veins with metal ores

There are numerous varieties of barite, including nitrobarite, radiobarite, and even fubarite. (The International Mineralogical Association recommends a different spelling—baryte—but it is rarely spelled that way in the United States.) Barite usually occurs as pale-yellow to brownish-white tabular crystals. Its color can range from white to dark brown, however, and even a striking blue, depending on impurities that substitute for the barium ion. Crystal structure in the field can range from slender prisms to fibrous masses. Barite has a white streak and a vitreous, glassy luster. Some barite specimens fluoresce, but after being heated to red hot and allowed to cool, all barite appears orange under a black light.

Barite is common throughout the Rockies, especially in Montana and Colorado. Striking blue barite crystals occur in weathered shale near Stoneham, Colorado. California also produces notable barite crystals, as do many central and eastern states.

Beryl

Common beryl, pale green but well formed with nice termination
PHOTO COURTESY OF THE RICE NORTHWEST MUSEUM OF ROCKS AND MINERALS

$Be_3Al_2Si_6O_{18}$
Family: Silicates
Mohs: 7½–8
Specific gravity: 2.7–2.9
Key test(s): Hard; hexagonal form
Likely locale(s): Granite pegmatites

Like corundum, beryl comes in a variety of formulas. Gemmy bright-green beryl is an emerald; gemmy blue beryl is aquamarine. When red or pink, beryl is morganite; other varieties include a golden beryl, which is yellow, and heliodor, which is greenish yellow. Goshenite is the name for colorless beryl. In all forms, beryl has a vitreous luster and a colorless streak. Crystals are hexagonal, sometimes in striking, perfectly six-sided prisms. Cleavage is not a good test, but hardness is, as beryl is harder than quartz. Common beryl is a constituent of many pegmatites in the Rockies, while gem-quality beryl typically occurs in vugs, cavities, and pockets.

Beryl occurs in Idaho's Sawtooth Mountains and in Montana near the Boulder Batholith. California's Pala District in San Diego County produces notable beryl. Colorado boasts numerous beryl outcrops, associated with pegmatites near Mount Antero and at Lake George. Look for beryl along the East Coast, especially associated with the Appalachians. North Carolina is especially productive.

Calcite

Large calcite crystals in a cluster

Calcium carbonate, CaCO$_3$
Family: Carbonates
Mohs: 3
Specific gravity: 2.7
Key test(s): Rhombohedral angles; hardness
Likely locale(s): Veins

Calcite is very common and is found as crystals in pegmatites, as veins in basalts, and mixed with zeolites, among limestone, and elsewhere. It is usually yellow but can occur in a pure, clear crystal form. It has a white streak, and its luster is vitreous but sometimes very dull. One of the classic field tests for calcite is the rhombohedral crystal habit, and calcite will cleave easily along three planes, which is distinctive as well. Calcite also fluoresces. Gypsum doesn't effervesce in acid. Two well-known crystal forms are dogtooth and nailhead calcite. Perfect calcite crystals display double refraction, in essence doubling the image they sit on.

Calcite specimens are common to igneous, metamorphic, and sedimentary rocks and are widespread throughout the United States. Most of the western states feature calcite deposits with good specimens. Tennessee and New Jersey both contain notable, collectible specimens, as do Kentucky, Arkansas, and Iowa.

Chalcedony

Soft white chalcedony has a waxy appearance.

Quartz, SiO$_2$
Family: Cryptocrystalline quartz
Mohs: 6½–7
Specific gravity: 2.6–2.64
Key test(s): No cleavage; conchoidal fracture
Likely locale(s): Quartz-rich areas

MINERALS

Chalcedony is one of the most common forms of quartz, and it is sometimes used as an all-inclusive term for a vast variety of collectible material, including agate, bloodstone, heliotrope, chrysoprase, jasper, and flint. When dark red or orange, chalcedony is called sard; when more translucent, it is called carnelian agate. Other varieties of chalcedony include layering, making for sardonyx; when found as green chalcedony with red spots it is called bloodstone, or heliotrope. Rockhounds call banded or clear chalcedony agate; if there are inclusions, the sample is moss agate. Apple-green chalcedony is chrysoprase. Opaque and red, yellow, or brown chalcedony is called jasper; when white or gray or dark gray, it is called flint. There is no cleavage, as there is no crystal habit, and the fracture is conchoidal. Chalcedony typically has a waxy, vitreous, or dull luster.

Chalcedony occurs in virtually all alluvial gravels to some degree. Its various forms are common throughout the United States.

Chrysocolla

Striking blue chrysocolla resembles turquoise.
PHOTO COURTESY OF THE RICE NORTHWEST MUSEUM OF ROCKS AND MINERALS

$(Cu,Al)_2H_2Si_2O_5(OH)_4 \cdot nH_2O$
Family: Silicate
Mohs: 2½–3½
Specific gravity: 1.9–2.4
Key test(s): Soft
Likely locale(s): Copper-producing regions

Chrysocolla is a pleasing blue or green material, usually banded, and softer than turquoise. It produces a streak that is slightly blue-green in color, which is helpful. Its luster is vitreous to dull, and it forms in botryoidal masses or coatings rather than as crystals. Chrysocolla oxidizes from copper deposits. Its softness precludes it from being an important lapidary material, but it will take a polish.

There are numerous chrysocolla deposits throughout the copper-mining regions of the Rockies, such as the Bay Horse and Seven Devils districts of Idaho. In Wyoming, search the copper-mining areas in Albany and Carbon Counties. Montana is a major copper-producing state, thus, the area around Butte is notewor-thy among hundreds of locales. Colorado also boasts numerous chrysocolla deposits.

Chrysotile

Fibrous chrysotile is an obvious source of asbestos fibers.
PHOTO COURTESY OF THE RICE NORTHWEST MUSEUM OF ROCKS AND MINERALS

$Mg_3(Si_2O_5)(OH)_4$
Family: Phyllosilicates
Mohs: 2½–3
Specific gravity: 2.53
Key test(s): Soft, silky
Likely locale(s): Serpentine belts

Chrysotile is the only asbestos from the serpentine family; all other asbestos minerals, such as tremolite and actinolite, are amphiboles. Known as the "white asbestos," chrysotile is the most common form and is typically composed of white, curly fibers. It has a silky luster and leaves a white streak. You may be able to scratch chrysotile, which is 2½ on the Mohs scale, with your fingernail. But you probably shouldn't, because then you'd get asbestos under your fingernail. The amphibole asbestos family forms thinner fibers and is more deadly to your lungs, but you should respect chrysotile as well.

Chrysotile occurs within the Green Mountain and State Line kimberlites in Colorado. Idaho lists limited chrysotile, such as the Hailey district. Montana hosts chrysotile in the Libby area, in the Stillwater Complex, and it is concentrated in Madison County. Wyoming hosts chrysotile among scattered asbestos deposits in Natrona and Converse Counties.

Corundum

Corundum in the form of very low-grade ruby (left) and sapphires (right) from Spokane Bar, Montana

Aluminum oxide, Al$_2$O$_3$
Family: Oxides
Mohs: 9
Specific gravity: 3.9–4.1
Key test(s): Only diamond is harder
Likely locale(s): Alluvial deposits

North American corundum comes in three main varieties: emery, which is black or dark gray and historically used for abrasives; sapphire, which can be blue, yellow, purple, or green; and ruby, which is red. Each variety has a hardness of 9 on the Mohs scale, so all forms of corundum scratch only by diamond. Luster ranges from vitreous all the way to adamantine, especially in true, gem-quality rubies and sapphires. The hexagonal crystal habit is not a good field indicator, mostly because it is rare to find crystals big enough to inspect. The hardness test is easier in any case.

Wyoming's Fremont County hosts a handful of corundum deposits; it is sparse elsewhere in the state. Idaho is similarly limited, with Burgdorf, Rock Flat, the Lochsa River, and Clearwater County prominent. Montana and Colorado are more noteworthy, with Montana topping the list and containing dozens of deposits. Emery is common in Pennsylvania.

Dioptase

Beautiful blue-green dioptase crystals are sometimes mistaken for emeralds.
PHOTO COURTESY OF THE RICE NORTHWEST MUSEUM OF ROCKS AND MINERALS

Copper silicate, $CuSiO_3H_2O$

Family: Copper silicates

Mohs: 5

Specific gravity: 3.3

Key test(s): Green streak

Likely locale(s): Oxidized copper sulfide zones

Dioptase is not a common mineral. It is a striking blue-green color and could be confused with emeralds if not for its softness—emeralds, a form of corundum, reach 8 on the Mohs scale; dioptase crystals are at 5. Dioptase has a green streak and vitreous luster and has perfect cleavage in three directions. It is often present as a thin, crystalline coating, so coarser specimens with distinguishable six-sided crystals fetch higher values. Such crystals are often brittle and fragile.

There are numerous dioptase occurrences throughout Arizona's copper-producing region; other locales include Clark County, Nevada, and San Bernardino County, California.

Epidote

Well-formed epidote crystals in the "jackstraw" habit, from Prince of Wales Island, Alaska
PHOTO COURTESY OF THE RICE NORTHWEST MUSEUM OF ROCKS AND MINERALS

$Ca_2Al_2FeOSiO_4Si_2O_7(OH)$
Family: Silicates
Mohs: 6
Specific gravity: 3.3–3.6
Key test(s): Color
Likely locale(s): Metamorphic rocks

Epidote is a common rock-forming mineral with a characteristic green color, although it can occur as yellow-green or even greenish black. Epidote has a colorless to gray streak. Crystals are common in the monoclinic system, forming long, slender crystals or massively tabular encrustations. Epidote occurs in many situations: in epidote schist; in granite pegmatites; in basalt cavities, along with andesite; in greenstones; and in regional metamorphic rocks.

Epidote is common across the Rocky Mountain states. Wyoming is probably the least represented, but Idaho, Montana, and Colorado have hundreds of known epidote locales. Several locales in Alaska are noteworthy.

Feldspars—Alkali—Microcline

Giant microcline crystals with black smoky quartz from the Sawtooth Mountains in Idaho
PHOTO COURTESY OF THE RICE NORTHWEST MUSEUM OF ROCKS AND MINERALS

KAlSi$_3$O$_8$
Family: Tectosilicates
Mohs: 6–6½
Specific gravity: 2.5–2.6
Key test(s): Hardness
Likely locale(s): Common rock-forming mineral

Alkali feldspars, sometimes known as potassium feldspars or potash feldspars, are usually white, light gray, or light pink. One exception is amazonite, which appears light blue or green. These varieties of microcline have a vitreous luster and leave a white streak. Feldspars are key rock-forming minerals for igneous and metamorphic rocks and make up more than 50 percent of the Earth's crust.

There are scattered locales across Idaho, with prized specimens coming from the Sawtooths. Montana also has numerous deposits, with good crystals in the Boulder Batholith, at Lemhi Pass, and elsewhere. Microcline occurs in numerous pegmatites in Wyoming, especially in Goshen and Albany Counties. Colorado has hundreds of microcline locales throughout the western and central regions. The California-Arizona border region contains numerous microcline locales; Arizona lists a handful, as do Nevada and New Mexico. The Zapot pegmatite in the Gillis Range of Mineral County, Nevada, is one of the only amazonite locales in the Southwest. Another locale is at Haystack Mountain in Inyo County, California.

MINERALS

Feldspars—Alkali—Microcline var. Amazonite

Beautiful blue amazonite crystals with black smoky quartz from Lake George, Colorado
PHOTO COURTESY OF THE RICE NORTHWEST MUSEUM OF ROCKS AND MINERALS

KAl AlSi$_3$O$_8$
Family: Tectosilicates
Mohs: 6–6½
Specific gravity: 2.5–2.6
Key test(s): Hardness
Likely locale(s): Pegmatites

Amazonite shares all the identification properties of microcline, but amazonite ranges from pale, light green to a deep, greenish blue. Prized specimens contain more color, as most amazonite is not as vivid as the museum-grade pieces. Amazonite occurs in large vugs and cavities within pegmatites, combining with smoky quartz and phenakite.

Colorado is the only state in the Rockies to produce amazonite. Pegmatites on private claims in the Lake George area yield very high-quality specimens. There are scattered amazonite deposits in California.

Feldspars—Plagioclase—Orthoclase

Orthoclase crystals in matrix
PHOTO COURTESY OF THE RICE NORTHWEST MUSEUM OF ROCKS AND MINERALS

Orthoclase: $KAl Si_3O_8$
Family: Tectosilicates
Mohs: 6–6½
Specific gravity: 2.6–2.8
Key test(s): Hardness
Likely locale(s): Common rock-forming mineral

Plagioclase feldspars rank by how much anorthite is present, ranging from 0 to 100 percent, from pure albite to pure anorthite. These common rock-forming minerals have a vitreous luster and a white streak. The color is variable but is typically dull white or light gray in appearance, with light tinting common. Feldspars show good cleavage in two directions at 90 degrees, and good twinning. Adularia is transparent. Sunstones are a form of oligo-clase feldspar. Labradorite is another gem variety of orthoclase feldspar.

Feldspars are common throughout the United States.

Fluorite

This vivid green fluorite is from Cheshire County, New Hampshire.
PHOTO COURTESY OF THE RICE NORTHWEST MUSEUM OF ROCKS AND MINERALS

Calcium fluoride, CaF$_2$

Family: Halide

Mohs: 4

Specific gravity: 3.0–3.2

Key test(s): Pastel purple, pink, green color; cubic crystal habit

Likely locale(s): Found in hydrothermal, sedimentary, igneous, and volcanic deposits

Fluorite comes in a variety of colors such as purple, green, and pink, and banding is common. It leaves a white streak, as do many minerals, so a better test is the hardness test or the square crystals. Fluorite crystals are unique in that they cleave perfectly in four directions. Fluorite has a vitreous luster and is softer than quartz but harder than calcite. Another good field test is to check for fluorescence. Fluorite veins are often associated with economic ore deposits, such as galena and sphalerite.

There are numerous fluorite locales across North America; one of the largest is in Newfoundland, Canada. One very good collecting site is a fee-dig at the Big Rock Candy Mountain in British Columbia. Idaho's famed Challis area offers decent fluorite samples, as do Meyers Cove and other scattered locales. Montana and Wyoming also offer limited fluorite collecting. The best state by far in the Rockies for collecting gem-quality fluorite is Colorado, especially in Jefferson, Boulder, and Teller Counties. New Hampshire, New Jersey, Virginia, and especially Illinois also host excellent fluorite locales.

Fulgurite

Crusty fulgurites form when lightning fuses quartz-rich sands.
PHOTO COURTESY OF THE RICE NORTHWEST MUSEUM OF ROCKS AND MINERALS

Quartz, SiO$_2$
Family: Quartz
Mohs: 7
Specific gravity: 2.65
Key test(s): Tubelike fused glass; rough exterior
Likely locale(s): Sandy deserts

Sand fulgurites are the result of lightning strikes hitting quartz-rich sand deposits. (Rock fulgurites result from lightning striking rocks, usually at the tops of mountains, and are less collectible.) The intense heat of the strike instantly fuses the quartz in the sand, usually in a tube that is rough on the outside and smooth, sometimes bubbly, on the inside. The resulting color is based on the melted material and is usually gray, light gray, or tan. Fulgurites can extend several feet into the ground, usually branching or eventually tapering away.

Look for fulgurites in any flat, open sandy desert or dune area where thunderstorms are common. This includes most of the southeast, southwest, and prairie states.

Garnet

Rough garnet from Emerald Creek, Idaho

Almandine, $X_3Y_2(SiO_4)_3$
Family: Metal silicates
Mohs: 6½–7½
Specific gravity: 3.6–4.3
Key test(s): Hackly fracture, hardness
Likely locale(s): Schist; black sands

Increasing metamorphism ⟶

Chlorite	Biotite	Garnet	Staurolite	Kyanite	Sillimanite

Taken as a family, garnet crystals show a vitreous luster, have no streak to speak of, and do not cleave. Garnet is harder than apatite, does not fluoresce like zircon, and has higher specific gravity than tourmaline, plus it is usually associated with schist. That rule isn't fixed, however, as some gem garnets are associated with

Pyrope	Deep red	$Mg_3Al_2(SiO_4)_3$
Almandine	Deep red to brown	$Fe^{2+}_3Al_2(SiO_4)_3$
Spessartine	Brown red	$Mn_3Al_2(SiO_4)_3$
Grossular	Colorless or light	$Ca_3Al_2(SiO_4)_3$
Andradite	Wine red or green	$Ca_3Fe^{3+}_2(SiO_4)_3$
Uvarovite	Emerald green	$Ca_3Cr_2(SiO_4)_3$

granitic pegmatites. Gem-quality garnet is somewhat rare; adding further value is a tendency for impurities to line up and create four-sided or six-sided stars.

Garnets are common across the United States. Wyoming has some scattered locales, mostly in Goshen County. Colorado and Montana have dozens of locales. Ruby Reservoir in Montana is an easy collecting spot. Some streams in northern Idaho run red with millions of tiny garnets plentiful enough to supply sandpaper companies with raw product.

Geodes

"Coconut" geode from Chihuahua, Mexico
PHOTO COURTESY OF THE RICE NORTHWEST MUSEUM OF ROCKS AND MINERALS

Quartz, SiO$_2$
Family: Silicate
Mohs: Varies
Specific gravity: Varies
Key test(s): Round shape
Likely locale(s): Rhyolite lava beds

Geodes are round structures that range greatly in size, shape, and filling. They differ from nodules and concretions in their formation—nodules and concretions occur in sedimentary rocks, while geodes are usually associated with lava beds. Some geodes are almost perfectly round, about the size of a small fist, and when completely full of quartz, agate, jasper, or chalcedony are usually called thundereggs. Some geodes are misshaped and hollow, but the most collectible specimens are lined with crystals.

Oregon's state rock is the thunderegg, with most coming from the fee-dig operation at Richardson's Ranch; there are lesser-quality deposits elsewhere in that state. Many states host geode locales, including Indiana, Iowa, Missouri, Kentucky, Utah, Idaho, Nevada, Colorado, Montana, and California. The Dugway geode beds in Utah and the Hauser geode beds near Wiley Well, California, are especially notable. The coconut geodes of Mexico also produce fine specimens.

Gypsum

Spar variety of gypsum
PHOTO COURTESY OF THE RICE NORTHWEST MUSEUM OF ROCKS AND MINERALS

Calcium sulfate, $CaSo_4 \cdot 2H_2O$
Family: Sulfate
Mohs: 1½–2
Specific gravity: 2.3–2.4
Key test(s): Softer than calcite
Likely locale(s): Hydrothermal basins

Gypsum is usually white in hand specimens but sometimes appears colorless in pure crystals. It leaves a white streak. Crystals are monoclinic; a rhombic pattern is common, as is the massive, fibrous "satin spar" form. The ram's horn variety is not common, but it is striking and worth noting. It can form as a "desert rose" with clusters of sheetlike "petals." Cleavage is perfect in one direction. The Mohs scale defines gypsum as a 2, although some varieties can be slightly softer. That makes it softer than calcite, which is an easy field test. Gypsum is common in hydrothermal replacement deposits, in veins, and as an evaporite in sedimentary basins, resulting in giant strip-mining operations for use in drywall or as fertilizer. One form of gypsum, selenite, was discovered in a Mexican cave in the form of nearly 40-foot crystals.

There are numerous gypsum deposits across the United States, especially Iowa, Texas, Utah, and New Mexico. The famed White Sands National Monument is composed of white gypsum particles.

Halite

Cubic halite showing excellent cleavage and vitreous luster
PHOTO COURTESY OF THE RICE NORTHWEST MUSEUM OF ROCKS AND MINERALS

Sodium chloride, NaCl
Family: Chlorides
Mohs: 2–2½
Specific gravity: 2.1–2.6
Key test(s): Taste
Likely locale(s): Evaporite deposits; sedimentary

Halite, or rock salt, is usually clear but is sometimes tinted pink or gray. It leaves a white streak. Halite is strongly isometric, and when pure it features well-formed crystals in tiny cubes. It has a vitreous luster and perfect cleavage in three directions. Halite is similar to cryolite but harder, and its salty taste makes for an easy field test. Halite typically occurs in large evaporite deposits and in large domelike underground structures. It mixes readily with other ions, and there are perhaps twenty halide minerals, such as sulphohalite and polyhalite.

Salt deposits are common in the western Basin and Range Province, principally from evaporated lakes. New York, Ohio, Kansas, and eastern Canada all host major underground halite deposits. Texas and Louisiana feature salt domes associated with oil and gas extraction. California's San Francisco Bay hosts colorful evaporation ponds used to harvest salt from seawater.

Hornblende

These coarse black hornblende crystals are larger than normal.

$Ca_2(Mg,Fe,Al)_5(Al,Si)_8O_{22}(OH)_2$

Family: Inosilicates

Mohs: 5–6

Specific gravity: 3.0–3.4

Key test(s): Cleavage angles

Likely locale(s): Common rock-forming mineral

Hornblende is a common rock-forming mineral found in granite, basalt, gneiss, gabbro, diorite, schist, and many other igneous and metamorphic rocks. It is a main constituent of an entire group of rocks known as amphibolites, all differing by their amounts of iron, magnesium, calcium, sodium, aluminum, and potassium. There is a rock known as hornblendite, but it is rare. Hornblende is usually black but sometimes green or brown. Hornblende leaves a colorless streak, and crystals are monoclinic, usually short, and prismatic. Cleavage is perfect in two directions. Hornblende can be confused with schorl, which is black tourmaline, but schorl does not cleave like hornblende. Diamond-shaped cleavage angles and a greenish-black color are the best field indicators.

Hornblende is rarely collectible as large crystals, but it is common in many mining districts.

Jasper

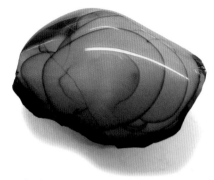

Polished picture jasper from Bruneau Canyon, Idaho

Quartz, SiO$_2$
Family: Cryptocrystalline quartz
Mohs: 7
Specific gravity: 2.65
Key test(s): Conchoidal fracture
Likely locale(s): Basalt flows

Jasper is another common form of quartz, derived from silica-rich ground solutions typically circulating through basalt, which technically makes it a form of chalcedony. Jasper thus differs from chert and flint because there is no association with biologic processes. Jasper comes in many shades and colors but is usually red, tan, or yellow and sometimes green. Gem-quality jasper is hard and shiny, while common jasper is typically porous and will not polish. Jasper occurs in most regions where basalt is common, but not always. It has no crystal habit and leaves a white streak. Native Americans frequently used jasper to fashion hand axes, arrowheads, and cutting tools.

Jasper is common in volcanic regions across the United States, especially in the Basin and Range Province.

Kyanite

Kyanite usually occurs as dark-blue blades.

$Al_2(SiO_4)O$
Family: Nesosilicate
Mohs: 5½–7
Specific gravity: 3.5–3.7
Key test(s): Blue, bladed crystals; splinters easily
Likely locale(s): Metamorphic terrain; pegmatites

Increasing metamorphism ⟶

Chlorite	Biotite	Garnet	Staurolite	Kyanite	Sillimanite

Kyanite is an indicator mineral that reveals an area has undergone intense regional metamorphism. It is typically blue, sometimes deep blue, and provides the coloring for blueschist. Kyanite is usually found in long, columnar crystals that appear fibrous or bladed. Crystals are often brittle. Kyanite is a great representative of an obscure mineralogical test known as anisotropism. Most minerals do not care how you measure their hardness, but kyanite is an exception. When you measure the hardness of kyanite, you will measure it at 7 if you go lengthwise along the blade, but only 5½ if you go across the short axis.

North Carolina produces exceptional gem-quality kyanite crystals suitable for faceting. Common kyanite is found in the metamorphic terrains of Idaho, Montana, Wyoming, and Colorado. Several eastern states contain kyanite schists; the Manhattan schist under New York City is a good example.

Mica Group—Biotite

Biotite sheets and flakes are usually quite dark.
PHOTO COURTESY OF THE RICE NORTHWEST MUSEUM OF ROCKS AND MINERALS

$K(Mg,Fe)_3(Al,Fe)Si_3O_{10}(OH,F)_2$
Family: Phyllosilicates
Mohs: 2–3
Specific gravity: 2.7–3.0
Key test(s): Platy cleavage
Likely locale(s): Schists

Increasing metamorphism ⟶

Chlorite	Biotite	Garnet	Staurolite	Kyanite	Sillimanite

The mica group is composed of several similar sheet-like minerals, including muscovite, biotite, phlogopite, lepidolite, and chlorite. Of the two most common varieties, muscovite is white or colorless; biotite is much darker. Mica experts recommend that the term *biotite* no longer be used for a specific chemical formula; instead they suggest using *biotite* as a field term for all dark micas. Well-formed micas occur in granitic pegmatites, a historic source of muscovite; phlogopite resides in marble and hornfels.

Dark black or brownish-black biotite micas are common throughout the United States, usually in schists and pegmatites.

Mica Group—Chlorite

This orthoclase feldspar is dusted with a coating of light green chlorite.
PHOTO COURTESY OF THE RICE NORTHWEST MUSEUM OF ROCKS AND MINERALS

$(Mg,Fe)_3(Si,Al)_4O_{10}(OH)_2 \cdot (Mg,Fe)_3(OH)_6$

Family: Phyllosilicates

Mohs: 2–2½

Specific gravity: 2.6–3.3

Key test(s): Green, soft; platy cleavage

Likely locale(s): Schists; hydrothermal ore deposits

Increasing metamorphism ⟶

Chlorite	Biotite	Garnet	Staurolite	Kyanite	Sillimanite

Like biotite, chlorite no longer refers to a single mineral. Instead it is the name for a group of related mica minerals, including clino-chlore, chamosite, cookeite, and several others. However, field-workers still use the term *chlorite* to refer to green coatings or stains on other minerals. Chlorite represents the lowest grade of metamorphism; it is a key constituent of greenschists and green-stones, for example. Chlorite has a white streak and is generally not collectible except for the rare forms.

Chlorite is common throughout the low-grade metamorphic rocks of the United States, often in major greenstone belts.

Mica Group—Muscovite

Muscovite sometimes occurs in large sheets called "books."
PHOTO COURTESY OF THE RICE NORTHWEST MUSEUM OF ROCKS AND MINERALS

$KAl_2(AlSi_3O_{10})(F,OH)_2$
Family: Phyllosilicates
Mohs: 2–2½ parallel to main cleavage, 4 at right angle
Specific gravity: 2.8–3.0
Key test(s): Platy cleavage; pearly luster
Likely locale(s): Schists

Increasing metamorphism ⟶

Chlorite	Biotite & Muscovite	Garnet	Staurolite	Kyanite	Sillimanite

Muscovite, the most common mica seen in the field, is usually white or colorless; biotite is much darker. Chlorite is greenish, while lepidolite is pink or light purple. Muscovite has a vitreous, pearly luster and leaves a colorless streak. Perhaps most characteristic, its cleavage is perfect in one direction, sometimes resulting in great sheets called "books." Because muscovite flakes can look yellow and shiny, they are sometimes referred to as fool's gold (as is pyrite), but a knifepoint will fracture pyrite and mica flakes, while flattening or cutting gold. If large enough, micas show a hexagonal pattern. Muscovite bends the easiest of the mica group.

Muscovite mica is common across the United States in metamorphic terrains.

Olivine

Olivine-rich basalt in the ground at Dunraven Pass, inside Yellowstone National Park

$(Mg,Fe)_2SiO_4$

Family: Nesosilicates

Mohs: 6½–7

Specific gravity: 3.5–4.0

Key test(s): White streak

Likely locale(s): Found with metamorphic rocks and in hydrothermal replacement deposits

Olivine is the name for a series of minerals, with forsterite on the magnesium-rich end of the series, and by far the most common, and fayalite on the iron-rich end. Olivine-group minerals are usually dark green, with a vitreous luster and colorless streak. When found in crystalline form, such as the semiprecious gemstone peridot, olivine will cleave in two directions, but it is more commonly associated with olivine-rich basalt.

Olivine-group minerals are very common. They're found in pallasite meteorites, composing much of the Earth's mantle, and concentrated at beaches in Hawaii that have turned green from the presence of so much olivine. Mafic rocks such as basalt and gabbro are rich in olivine; ultramafic rocks such as peridotite and dunite also contain abundant olivine.

Onyx

Sliced and polished onyx, with characteristic banding
PHOTO COURTESY OF THE RICE NORTHWEST MUSEUM OF ROCKS AND MINERALS

Quartz, SiO$_2$
Family: Quartz
Mohs: 6–7
Specific gravity: 2.6–2.7
Key test(s): Patterns
Likely locale(s): Silica-rich environments

Onyx is yet another form of the chalcedony family of quartz. Look for uniform banding, creating striking patterns of red, white, black, or brown. If the bands contain sard, the result is called sardonyx. The most common color is black, but rockhounds often label as onyx material formed in white carbonates such as limestone, marble, and calcite. Mineralogists restrict onyx to a variety of agate. The luster is vitreous to silky, and the streak is white; onyx is usually slightly translucent.

Onyx is common in limestone caverns across the United States. West Virginia, Kentucky, Michigan, New York, Iowa, Montana, California, New Mexico, and Washington all have onyx deposits, as do many other states.

Opal

Common opal showing banding

Quartz, $SiO_2 \cdot nH_2O$
Family: Cryptocrystalline quartz
Mohs: 5½–6½
Specific gravity: 2.0–2.2
Key test(s): Softness, lighter than other quartz
Likely locale(s): Silica-rich environments

Opal is another form of quartz, but this time with considerable water present. There are numerous varieties of common opal, with mindat.org showing at least 155 names. Common opal, sometimes referred to as hyalite, has a distinctive vitreous, pearly luster. It is usually white, but colors typically include yellow, brown, green, blue, or gray. Its streak is white. Opal is not crystalline; there is no habit or cleavage, and it has a conchoidal fracture. Opal is softer than agate, jasper, and other forms of quartz. Opal frequently forms in cavities, fractures, and air bubbles in basalt, where it sometimes weathers out and remains as small, round pea-size blebs. Wood and shells often become "opalized" through replacement. Opal will fluoresce, which also sets it apart from jasper and quartz.

Common opal occurs in many states across the country.

Orpiment and Realgar

Yellow orpiment (left) and red realgar (right)
PHOTOS COURTESY OF THE RICE NORTHWEST MUSEUM OF ROCKS AND MINERALS

As_2O_3 and As_4S_4
Family: Arsenides
Mohs: 1½–2
Specific gravity: 3.5–3.6
Key test(s): Orpiment is bright yellow; realgar is deep red
Likely locale(s): Hot springs and fumaroles

These two arsenic minerals usually show up with each other; orpiment is an oxide of arsenic, and realgar is a sulfide. Orpiment has a bright yellow streak; ancient artists and painters used orpiment as a yellow pigment. Realgar is deep orange-red. Orpiment is often a by-product of weathered, altered realgar, where the sulfur has leached away, leaving an oxide. Likely collecting locales include hot springs and hydrothermal vein systems. Orpiment rarely occurs as a crystal and is much more common as a fibrous, velvety coating, sometimes in the botryoidal habit. Realgar often displays as small, granular crystals; larger specimens are highly sought.

Orpiment and realgar are found in many western states, as well as New York, New Hampshire, and Pennsylvania. North Carolina also has a notable realgar deposit in Swain County.

Phenakite

Water-clear rhombohedral phenakite specimen with excellent crystal faces
PHOTO COURTESY OF THE RICE NORTHWEST MUSEUM OF ROCKS AND MINERALS

Be_2SiO_4
Family: Nesosilicates
Mohs: 7½–8
Specific gravity: 2.96
Key test(s): Hardness
Likely locale(s): Pegmatites

Phenakite occurs in a wide variety of disguises. It can be tabular, it can be columnar, and it can even occur in granular sheets. It can be colorless, white, gray, or even pale pink. It has a striking, colorless crystal that is occasionally gemmy enough to facet. Phenakite is harder than 7½ and thus will scratch quartz. Phenakite and beryl are equally hard, but beryl is hexagonal and usually forms longer crystals.

Phenakite occurs in several western states, especially Idaho and Colorado. Colorado contains striking rhombohedral phenakite, in association with amazonite in the pegmatites near Lake George in Park County. More-elongated phenakite crystals originate in Colorado's Mount Antero region. Many New England states host phenakite crystals in pegmatite.

Quartz

Straight, clear quartz showing excellent crystal faces and pointed termination

Crystalline quartz, SiO$_2$

Family: Silicates

Mohs: 7

Specific gravity: 2.65

Key test(s): Hardness; glossy luster

Likely locale(s): Pegmatites for gem material

After feldspar, quartz is the most dominant mineral in the Earth's crust. It is a semiprecious gem in multiple forms, such as purple amethyst, clear agate, yellow citrine, pink rose quartz, orange carnelian, and black smoky quartz. The hardness is 7; there is a white streak if you can produce it; and crystals are hexagonal, with no cleavage. Collectors prize the specimens with a characteristic pointed, pyramidal termination. The black coloring of smoky quartz appears to be the result of free silicon atoms irradiated by the natural background radiation common to granite.

Because there are so many pegmatites in the United States, gem-quality quartz is abundant in many regions. Herkimer "diamonds" of New York are doubly terminated and water clear. North Carolina's Spruce Pine Gem Mine produces extremely pure quartz; Arkansas also produces exceptional specimens.

Rhodochrosite

The Alma Rose, one of the finest rhodochrosite specimens in the world, is from the Sweet Home Mine in Colorado.
PHOTO COURTESY OF THE RICE NORTHWEST MUSEUM OF ROCKS AND MINERALS

Manganese carbonate, MnCO$_3$
Family: Carbonate
Mohs: 3½–4
Specific gravity: 3.7
Key test(s): Deep pink to rose red; white streak, vitreous luster
Likely locale(s): Vein mineral in silver districts

Rhodochrosite is a lovely rosy red in its pure form, but impurities typically temper that vivid color down to pink, cinnamon, or even yellow. Its rhombohedral crystals are rare and highly collectible, but most collectors encounter field deposits as ribbons or banded minerals. These specimens will often take a polish and produce striking red-pink, banded lapidary creations. As crystals, rhodochrosite is soft and difficult to fashion into jewelry; it has perfect cleavage, another strike against it for faceting.

Rhodochrosite is fairly common in massive form across the western United States. New England, Pennsylvania, North Carolina, Arkansas, and Oklahoma also contain deposits. Alma, Colorado, boasts some of the best rhodochrosite in the world, including the Alma Rose, Alma King, and Alma Queen, all found in a plate of beautiful deep red crystals.

Rhodonite

Light pink color is characteristic of rhodonite, as are the dark manganese "roads" through the pink material.
PHOTO COURTESY OF THE RICE NORTHWEST MUSEUM OF ROCKS AND MINERALS

Manganese silicate, MnSiO$_3$
Family: Metal silicates
Mohs: 5½–6½
Specific gravity: 3.6–3.8
Key test(s): Pink (when fresh); turns brown
Likely locale(s): Metamorphic rocks; hydrothermal replacement deposits

Rhodonite is usually easy to identify because it is rose pink in the field and is one of the few minerals found in that color. Some rhodonite samples are red, brown, or yellow, depending on the calcium, iron, or magnesium impurities present with the manganese, so the color test is not foolproof. Most rhodonite samples are massive and pink, with dendrites of manganese-rich psilomelane that look like tiny roads. You can use that roadlike black tracking as a memory aid—roads in rhodonite.

There are rhodonite deposits across western US mining districts, including Oregon, Montana, California, Arizona, and Colorado. Michigan, Virginia, New York, and Maine also host rhodonite locales.

Selenite

Splendid selenite crystals from Cave of Swords in Chihuahua, Mexico
PHOTO COURTESY OF THE RICE NORTHWEST MUSEUM OF ROCKS AND MINERALS

$CaSO_4 \cdot 2H_2O$
Family: Sulfates
Mohs: 2
Specific gravity: 2.3
Key test(s): Yellow to water-clear crystals
Likely locale(s): With gypsum

Selenite is a form of gypsum and is a general term for satin spar, desert rose, gypsum flower, and selenite crystals. It does not contain the element selenium. Rockhounds typically reserve the name *selenite* for superior, transparent gypsum crystals found either individually or as masses. Selenite comes in a variety of colors, including brownish green, brownish yellow, grayish white, and other light tints. Primarily, however, think of the most water-clear tabular and columnar crystals as being selenite. Selenite has a white streak and a pearly luster.

Most western states host selenite deposits, including Idaho, Wyoming, California, and Arizona. Colorado has multiple locales, including in San Juan and El Paso Counties. New England, Virginia, Arkansas, Missouri, and Michigan also contain selenite deposits, as do many other states.

Sillimanite

Various rounded sillimanite pebbles. This mineral is difficult to locate as pure crystals.
PHOTO COURTESY OF THE RICE NORTHWEST MUSEUM OF ROCKS AND MINERALS

Al_2SiO_5
Family: Nesosilicates
Mohs: 7
Specific gravity: 3.2
Key test(s): Silky, fibrous
Likely locale(s): Severely metamorphosed aluminum-rich sedimentary rocks

Increasing metamorphism ⟶

Chlorite	Biotite	Garnet	Staurolite	Kyanite	Sillimanite

Sillimanite is an indicator mineral for rocks that have undergone metamorphism at the highest temperature. In other words, rocks with sillimanite got hotter than rocks with kyanite—even though the two minerals share the same formula (along with andalusite). Sillimanite crystals are usually rare and small; it is more common to find rocks with lots of sillimanite present, and they sometimes appear to shimmer. This is due to the silky luster, the fibrous mass of crystals, and the wavy nature of their alignment.

Idaho, Colorado, and California contain notable sillimanite deposits, as do Colorado and South Dakota. The metamorphosed terrains of the eastern United States, including New England, Pennsylvania, North Carolina, and Georgia, are also important.

Staurolite

Staurolite twins are a collectible form of the mineral.
PHOTO COURTESY OF THE RICE NORTHWEST MUSEUM OF ROCKS AND MINERALS

$Fe_2Al_9Si_4O_{22}(OH)_2$
Family: Nesosilicates
Mohs: 7–7½
Specific gravity: 3.7–3.8
Key test(s): Hardness, twinning
Likely locale(s): Highly metamorphosed schists

Increasing metamorphism ⟶

Chlorite	Biotite	Garnet	Staurolite	Kyanite	Sillimanite

Staurolite, which mainly occurs in staurolite schist, is an indicator mineral for intense metamorphism. A staurolite schist denotes a regionally metamorphosed rock between garnet and kyanite grade. Staurolite crystals are usually brown, with a vitreous luster and a white streak. The crystal habit is monoclinic; twinning, such as right-angle "fairy" crosses, is common. Cleavage is poor, lengthwise in direction.

Many of the Rocky Mountain states contain staurolite schist, but the collectible crosses are notable in older rocks of Virginia, Pennsylvania, and along the East Coast. Staurolite is the state gem of Georgia.

Talc

Talc is so soft, it is easy to carve with your fingernail.
PHOTO COURTESY OF THE RICE NORTHWEST MUSEUM OF ROCKS AND MINERALS

$Mg_3Si_4O_{10}(OH)_2$
Family: Magnesium silicates
Mohs: 1
Specific gravity: 2.7–2.8
Key test(s): Soft
Likely locale(s): Metamorphic terrain

Talc is easy to test in the field because you can scratch it with your fingernail—it represents 1 on the Mohs hardness scale. It is the main component of most soapstones and usually occurs as a mix of magnesium silicates, such as tremolite or magnesite. Beware of the tendency for asbestos to accompany talc, and avoid breathing its dust. Color varies, depending on associated minerals present, and can range from white to bright green. The streak is white. Talc has a greasy feel and shows a pearly luster. Crystals are extremely rare. Look for talc in metamorphic regions with ultramafic rocks such as dunite or peridotite.

Talc is common across the United States, especially in western states and along the East Coast.

Tourmaline—Schorl

Large, black, and chunky crystal of schorl, the common form of tourmaline

$Na(Mg,Fe)_3Al_6(BO_3)_3(Si_6O_{18})(OH,F)_4$

Family: Cyclosilicates
Mohs: 7–7½
Specific gravity: 3.0–3.3
Key test(s): Black; striations; harder than apatite
Likely locale(s): Pegmatites

Tourmaline comes in a variety of forms, with the most common being black schorl. Semiprecious gem varieties include elbaite, indicolite, rubellite, and dravite. Tourmaline has a vitreous luster, leaves a white streak, and forms a hexagonal crystal with no cleavage. Hornblende is also black, but schorl is harder and has a triangular cross section and is frequently striated, or lined.

Most western states that host pegmatites contain schorl. Along the East Coast, schorl occurs in Georgia, North Carolina, Virginia, Pennsylvania, and throughout New England.

MINERALS

Variscite

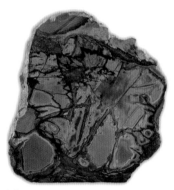

Variscite specimen from the Turquoise Mountain Mine in Fremont County, Idaho
PHOTO COURTESY OF THE RICE NORTHWEST MUSEUM OF ROCKS AND MINERALS

$AlPO_4 \cdot 2H_2O$
Family: Phosphate
Mohs: 4½
Specific gravity: 2.6
Key test(s): Greenish; veined
Likely locale(s): Phosphate regions

Variscite is a rare secondary mineral formed when phosphate-rich groundwater leaches through aluminum-rich rocks. It usually forms as attractive greenish-blue masses or nodules, frequently shot through with white, brown, or black veins. It is lighter than malachite and greener than turquoise, two minerals with which it can be confused. Variscite has a white streak and a vitreous-to-waxy luster, and it takes a nice polish. There are many known varieties of this mineral, depending on the ratio of aluminum and iron, the presence of arsenic, and the amount of other elements available.

Variscite notably occurs in California, Idaho, Nevada, Utah, and Arizona in the western United States. There are multiple locales in Arkansas, Georgia, North Carolina, Virginia, and Pennsylvania, as well as other states.

Zeolite Group—Heulandite

Nice pink heulandite specimen from the Rat's Nest Mine, near Challis, Idaho
PHOTO COURTESY OF THE RICE NORTHWEST MUSEUM OF ROCKS AND MINERALS

$(Ca,Na)(Al_2Si_7O_{18}) \cdot 6H_2O$

Family: Zeolites

Mohs: 3–3½

Specific gravity: 2.2

Key test(s): Pink (when fresh)

Likely locale(s): Basalt flows and hornfels

Heulandite is the name of a group of zeolites known for a distinctive pink-colored form, but crystals are often white or lack color completely. The streak is white. Crystals are monoclinic and sometimes display a unique coffin shape. Luster can be pearly or vitreous. Heulandite is usually found with other zeolites in veins or pockets within basalt and andesite but can also occur in schists.

Heulandite is common in volcanic regions across Oregon, Washington, Idaho, California, and Colorado, among other western states. Along the East Coast, heulandite occurs in New England, Pennsylvania, Arkansas, Virginia, and North Carolina.

MINERALS

105

Zeolite Group—Mordenite

Superb mordenite, said to be the world's finest, from the Rat's Nest Mine near Challis, Idaho
PHOTO COURTESY OF THE RICE NORTHWEST MUSEUM OF ROCKS AND MINERALS

$(Ca, Na_2, K_2)Al_2Si_{10}O_{24}\cdot7H_2O$
Family: Zeolites
Mohs: 3–4
Specific gravity: 2.1
Key test(s): Fibrous
Likely locale(s): Veins and cavities in basalt flows

Mordenite often forms sprays of delicate needles, which makes it one of the easier zeolites to identify, although the zeolite mesolite also occurs in needles. Mordenite can also form round balls. It is usually white, although it can discolor quickly if it begins to absorb moisture. It streaks white or colorless and usually forms as needlelike crystals but can also form velvety coatings. It usually occurs in the cavities and vugs of volcanic rocks.

Challis, Idaho, supplies some of the best mordenite specimens found anywhere. Other western states with mordenite sites include Oregon, Washington, California, and Arizona. There are scattered locales in New England.

Zeolite Group—Stilbite

Stilbite crystals are usually short, stubby, and milky white.
PHOTO COURTESY OF THE RICE NORTHWEST MUSEUM OF ROCKS AND MINERALS

$NaCa_2Al_3Si_{13}O_{36} \cdot 16H_2O$

Family: Zeolites

Mohs: 3½–4

Specific gravity: 2.12–2.22

Key test(s): Pearly, vitreous luster

Likely locale(s): Basalt flows and hornfels

Like heulandite, *stilbite* is the name for a series of zeolites rather than a single species; the difference is in the ratio of sodium and calcium present. Most stilbites are richer in calcium. Crystals are usually colorless or white, but they can be pink on occasion. Iceland has long been a prime resource for excellent stilbite specimens, but stilbite is a common zeolite and occurs alongside the rest of the zeolite family in vugs and cavities within zeolite-rich basalts.

Stilbite is common in many zones of the Columbia River basalts and related rocks of the Pacific Northwest. Many western states contain stilbite, among other zeolites, in volcanic zones. Several eastern states, including Georgia and Virginia, also host stilbite locales.

Metallic Minerals
Arsenopyrite

Arsenopyrite crystals have a warped appearance compared to pyrite.
PHOTO COURTESY OF THE RICE NORTHWEST MUSEUM OF ROCKS AND MINERALS

FeAsS
Family: Metal sulfide
Mohs: 5½–6
Specific gravity: 5.9–6.2
Key test(s): Density; garlic odor when crushed
Likely locale(s): Pegmatites, vein systems

Arsenopyrite is a variation on common pyrite, with an arsenic ion substituting for iron in the crystal lattice. That substitution tends to warp the crystals—where pyrite has more striking angles and lines, arsenopyrite is frequently striated and twinned. While pyrite is a yellow brass color, arsenopyrite is often more of a silver-white or gray. There are at least five other variations when other ions substitute for the arsenic ion, with cobalt, nickel, stibnite, gold, iridium, and others present in varying ratios. Arsenopyrite is not very stable and tends to oxidize; thus, it is one of the culprits blamed for the reddish-orange stains draining from sulfide mines. Interestingly, this variant on fool's gold can be a key indicator to the presence of large amounts of gold, and it is associated worldwide with tin and nickel deposits. The bad news is that arsenopyrite is considered "refractory"—meaning it is difficult to recover the gold.

Arsenopyrite is common in the metal deposits of the West. Other notable deposits occur in Georgia, North Carolina, Virginia, and New England.

Azurite

Velvety blue azurite from the Bisbee area of Arizona
PHOTO COURTESY OF THE RICE NORTHWEST MUSEUM OF ROCKS AND MINERALS

Copper carbonate, $Cu_3(CO_3)_2(OH)_2$
Family: Metal carbonate
Mohs: 5½–6
Specific gravity: 3.5–4.0
Key test(s): Color; hardness
Likely locale(s): Rich copper deposits

Azurite is a striking blue mineral closely associated with green malachite in copper-mining regions. The two make a pleasing combination when fresh, but azurite tends to lose its brilliance if subjected to heat, bright light, or simply too much air over too long a time. When fresh it has a bright-blue streak, which artists exploited for paint pigments in Europe during the Middle Ages. Color and streak are the two key characteristics when identifying field samples. Actual crystals are beautiful but rare; azurite tends to show up as velvety coatings.

Azurite is widespread in western US copper districts, especially in Arizona's Bisbee area. The Leadville district of Colorado is also noteworthy. Other locales include Mexico, British Columbia, Alaska, and along the East Coast in mineralized areas.

MINERALS

109

Bornite

Rockhounds sometimes refer to bornite as the "peacock ore" due to its many colors.
PHOTO COURTESY OF THE RICE NORTHWEST MUSEUM OF ROCKS AND MINERALS

Copper iron sulfide, Cu_5FeS_4
Family: Metal sulfides
Mohs: 3–3¼
Specific gravity: 4.9–5.3
Key test(s): Peacock color; heft
Likely locale(s): Rich copper deposits

Rockhounds call bornite the "peacock ore" because it has so many red, purple, bronze, blue, and other hues. It is an important sulfide of copper because it is rich in copper by weight. Bornite has a metallic luster, and its streak is grayish black, but you probably won't need to streak it for identification purposes, thanks to the striking play of colors. Bornite crystals are rare, as they generally appear as disseminated deposits in skarns, veins, and massive sulfide deposits.

Bornite is a principal copper ore throughout the western United States and is associated with copper deposits along the East Coast, in Michigan and Minnesota, in Missouri, and elsewhere.

Chalcocite-Digenite

Chalcocite tarnishes quickly and is usually associated with bornite.
PHOTO COURTESY OF THE RICE NORTHWEST MUSEUM OF ROCKS AND MINERALS

Copper sulfide, Cu_2S
Family: Metal sulfide
Mohs: 2½–3
Specific gravity: 5.5–5.8
Key test(s): Dense; black to lead-gray streak
Likely locale(s): Known copper areas

Chalcocite is another important copper ore. It rarely occurs as crystals, but when it does, it forms typically in hydrothermal vein systems, with a metallic luster and in tabular form. Chalcocite is much more common as a secondary mineral in massive oxidized zones, where the copper has leached out from other minerals. There are at least nine other copper-sulfide minerals that have varying ratios of copper and sulfur; these compose the Chalcocite-Digenite group.

Chalcocite is common in most US copper districts. Notable specimens come from the Bristol Copper Mine in Connecticut.

Chalcopyrite

Cluster of massive chalcopyrite crystals
PHOTO COURTESY OF THE RICE NORTHWEST MUSEUM OF ROCKS AND MINERALS

Copper iron sulfide, CuFeS$_2$
Family: Metal sulfide
Mohs: 3½–4
Specific gravity: 4.1–4.3
Key test(s): Crystals less cubic than pyrite
Likely locale(s): Known copper areas

Chalcopyrite is a common sulfide that is chemically quite similar to common pyrite, except chalcopyrite contains both copper and iron, whereas pyrite has no copper. Like pyrite, chalcopyrite is brassy and golden yellow, but chalcopyrite is softer than pyrite, and its crystal habit is a tetrahedron, while pyrite's is cubic. The streak is greenish black, also similar to pyrite. There is an entire family of minerals related to chalcopyrite—for example, with varying amounts of silver and gallium substituting for iron or copper and selenium substituting for sulfur. Chalcopyrite often occurs with hydrothermal economic ore deposits that host silver and gold. It forms massive sulfide zones with other sulfides, especially pyrite, and is a primary copper ore.

Most of the metal sulfide mines and prospects in the United States that contain pyrite also contain chalcopyrite. There are numerous locales across the Rockies and throughout the Southwest.

Cinnabar

Sugary coating of pale red cinnabar
PHOTO COURTESY OF THE RICE NORTHWEST MUSEUM OF ROCKS AND MINERALS

Mercury sulfide, HgS
Family: Metal sulfide
Mohs: 5½–6
Specific gravity: 3.5–4.0
Key test(s): Crimson, pink (when fresh); hard
Likely locale(s): Calderas, limestones

Cinnabar is easy to identify in the field, being a vivid red. Hexagonal crystals are quite rare; instead you should look for reddish streaks and crusts in sulfide-rich zones. If there is enough of a sample for a streak test, cinnabar leaves a scarlet streak. In rich mercury mines with heavy concentrations of cinnabar, miners have actually noted liquid mercury deposits in small pools and puddles, so look for that. As anyone who played with a broken thermometer as a child can attest, mercury is liquid at room temperature, but the vapors are dangerous. Cinnabar is heavy and bright red, but it can be confused with ochre, rust, and even common red paint (personal experience). Cinnabar samples tend to be heavier than ochre, and the streak is more vivid than hematite. When cinnabar appears agatized or opalized, rockhounds use the term *myrickite*.

Cinnabar is common across the western United States, especially in Washington, Oregon, and California. Arizona, Utah, and Arkansas also host good deposits.

Copper

Large crystalline copper mass from Michigan
PHOTO COURTESY OF THE RICE NORTHWEST MUSEUM OF ROCKS AND MINERALS

Copper, Cu
Family: Elemental metal
Mohs: 2½–3
Specific gravity: 8.9
Key test(s): Shiny as a new penny; heavy
Likely locale(s): Found with metamorphic rocks and in hydrothermal replacement deposits

When fresh, native copper can appear as bright as a shiny new penny. However, copper quickly tarnishes when exposed to air and starts to turn black, green, or blue. Native copper nuggets leave a characteristic copper streak and are easy to identify without much practice, thanks to the copper coins in our financial system. Crystals are rare and are in the isometric habit. Copper is very malleable, meaning it bends and can be pounded, flattened, or rolled.

Michigan's Keweenaw Peninsula is a premier area to look for large copper nuggets. Major copper mines still dot the western United States, although native copper is scarce. Idaho contains numerous native copper deposits, including good specimens from the Blackbird Mine in Lemhi County. Wyoming also has multiple copper occurrences, notably in Albany County. Colorado's Montrose County contains nice copper specimens. Montana's Butte district is particularly notable in that state. Arizona's Bisbee district still produces notable copper specimens.

Cuprite

Cuprite has a reddish tint and is very heavy.
PHOTO COURTESY OF THE RICE NORTHWEST MUSEUM OF ROCKS AND MINERALS

Cu_2O
Family: Oxides
Mohs: 3½–4
Specific gravity: 6.1
Key test(s): Dark-red crystals; streak
Likely locale(s): Copper regions

Cuprite forms impressive dark red crystals in the cubic habit, but few locales ever yield crystals large enough to facet. More common are cuprite hairs, which are fragile and unsuitable for lapidary work; even large crystals would be difficult to facet, being soft. Luster varies, from submetallic to a sharp, adamantine appearance. Penetration twins are common, and twelve-sided dodecahedrons are rare. The streak is unusual, appearing as a shiny, metallic red or metallic red-brown.

There are numerous cuprite locales associated with the bigger copper districts of the West.

Enargite

Enargite with pyrite, from the Leonard Mine in Silver Bow County, Montana
PHOTO COURTESY OF THE RICE NORTHWEST MUSEUM OF ROCKS AND MINERALS

Cu_3AsS_4
Family: Sulfides
Mohs: 3
Specific gravity: 4.4
Key test(s): Distinct cleavage, metallic luster
Likely locale(s): Hydrothermal veins

Enargite typically occurs as small, tabular crystals or black, sooty masses in sulfide veins. Enargite shares similarities with several other minerals, with substitution for zinc, silver, antimony, arsenic, and others. It is often associated with quartz and pyrite and makes attractive display pieces in crystal form. Most times, however, crystals are quite small.

Enargite is common in Colorado, Arizona, Missouri, and Utah; other states have small deposits. The premier collecting locale for enargite is in Silver Bow County, Montana.

Galena

Note the stair-step crystal pattern and dull, lead color of this galena specimen from Missouri.
PHOTO COURTESY OF THE RICE NORTHWEST MUSEUM OF ROCKS AND MINERALS

Lead sulfide, PbS
Family: Metal sulfide
Mohs: 2½
Specific gravity: 7.4–7.6
Key test(s): Gray, soft, stairways
Likely locale(s): Known sulfide areas

Galena is the primary ore for lead, and thus specific gravity is one of the key clues for identifying galena in the field. The drab, bluish-gray color; dull luster; and cubic, stair-stepped crystal structure are other easy clues to spot. Finally, galena is relatively soft; you should be able to scratch it with your fingernail or a piece of calcite. Galena leaves a dark gray streak, which is distinctive once you've seen it. It has a brittle fracture and can look like stibnite, the ore of antimony, but stibnite is harder. Cubes are common and unforgettable as well. Galena is usually found in veins and larger disseminated deposits and can mix with argentite, tetrahedrite, or sphalerite to create rich lead, silver, and zinc ores.

Galena was a common metal sulfide across the West. Idaho's famed Silver Valley produced excellent galena specimens in the past as a by-product of silver mining. Lead districts in Kansas, Missouri, and Arkansas produce excellent specimens.

Goethite

As goethite continues to oxidize, it takes on a yellow or orange pigment.
PHOTO COURTESY OF THE RICE NORTHWEST MUSEUM OF ROCKS AND MINERALS

FeO(OH)
Family: Iron oxide
Mohs: 5–5½
Specific gravity: 3.3–4.3
Key test(s): Crumbly; iron staining
Likely locale(s): Weathered iron mineralization

Goethite is the mineral residue left behind when an iron sulfide such as pyrite loses its sulfur ion. Because it is rich in iron, it serves as a valuable iron ore. Ancient cave painters used goethite for pigments, calling it ochre. Goethite often occurs with limonite, another rusty yellow iron ore. It is typically brown, yellow, or even orange, usually as a mass but sometimes as a coating. The streak shows the same variations. Crystals are rare, but goethite easily forms pseudomorphs as it oxidizes from a sulfur-rich parent. It is not a handsome mineral, nor is it generally collectible, but it is common and worth learning to recognize.

Goethite is scattered across the United States in every major metal sulfide district.

Gold

Gold foil on quartz from Colorado
PHOTO COURTESY OF THE RICE NORTHWEST MUSEUM OF ROCKS AND MINERALS

Au
Family: Elemental metal
Mohs: 2½–3
Specific gravity: 19.3
Key test(s): Weight, color
Likely locale(s): Quartz veins

Gold is very dense, at around 19 grams per cubic centimeter, but it is also soft enough to scratch with calcite. Those two keys alone would be enough if you could find a big enough specimen to test. Instead, most prospectors see tiny specks of gold in their pans, where neither test is practical. Two minerals have been called "fool's gold" because they mimic gold's appearance—pyrite, which is brassy, harder, and smells like sulfur when crushed; and mica, which breaks easily with a knifepoint and tends to float. Gold crystals are exceedingly rare, but they are isometric with no cleavage. The density is the best clue, however. Prospectors wash gold from gravels using tools such as pans, sluices, dredges, or high-bankers when there is enough water; in the desert they use dry washers.

Gold was a major draw for miners opening up the United States, and few nuggets remain. From Dahlonega, Georgia, to Nome, Alaska, the promise of gold unlocked wilderness areas and drove civilization. California is the most obvious place to start a serious quest, but Colorado, Montana, Oregon, Washington, and Idaho all deserve mention.

Hematite

Shiny, polished hematite showing the round, pillowy "kidney stone" habit
PHOTO COURTESY OF THE RICE NORTHWEST MUSEUM OF ROCKS AND MINERALS

Iron oxide, Fe_2O_3
Family: Oxides
Mohs: 5–6
Specific gravity: 4.9–5.3
Key test(s): Bright red on metal
Likely locale(s): Iron-rich mineralized areas

Hematite is rich in iron and is the most common iron ore, thanks to massive banded iron formations. In the field hematite can appear gray or black and has a distinct metallic luster, but after prolonged exposure to air, hematite will eventually start to show its characteristic rusty-red signature. There are multiple variations of hematite, depending on what ions have substituted for iron in the crystal lattice. Hematite has a striking red streak, which is one telltale sign. Crystals are varied; they can be hexagonal, tabular, or columnar. Hematite can also display a rounded, bubbly botryoidal habit. Hematite powder makes up red ochre, one of the oldest color tints known to man. Yellow ochre is also hematite, but it contains extra water that results in a yellow, not red, color.

Hematite is common in most US mining districts but is especially prominent in the Mesabi iron mines of Minnesota and the South Pass district of Wyoming.

Magnetite

Magnetite in crystalline form from the Spring Mountain district of Lemhi County, Idaho
PHOTO COURTESY OF THE RICE NORTHWEST MUSEUM OF ROCKS AND MINERALS

Iron oxide, Fe_3O_4
Family: Metal oxide
Mohs: 5½–6½
Specific gravity: 4.9–5.2
Key test(s): Magnetic; heavy, hard, and dark; attracts magnetic dust
Likely locale(s): Iron-rich mineralized areas

Magnetite, also known as lodestone, is easy to identify in the field thanks to its magnetic properties. It is usually gray-black to dark black, with a metallic luster, and leaves a black streak. Crystals are rare and usually small in the isometric habit and typically form stubby, doubly terminated octahedrons. Magnetite is quite common, as it tends to remain behind when many igneous rocks erode. Almost all rivers and streams carry some quantity of magnetite-rich "black sands" in their cracks and underneath bigger rocks. Gold panners report that there is always black sand with placer gold, but there is not always visible gold in black sands. Many prospectors save all their black sands because, in addition to magnetite, they can usually find palladium, platinum, and other rare metals in the mix. Ocean beaches typically concentrate magnetite, thanks to wave action.

There are thousands of magnetite showings in the United States.

Malachite

Unpolished malachite showing botryoidal habit
PHOTO COURTESY OF THE RICE NORTHWEST MUSEUM OF ROCKS AND MINERALS

Copper carbonate, $Cu_2CO_3(OH)_2$
Family: Metal silicates
Mohs: 3½–4
Specific gravity: 4.1–4.3
Key test(s): Green; presence of azurite
Likely locale(s): Low-grade copper deposits

This mineral can be a welcome sign in otherwise perplexing or barren outcrops. Even in low concentrations, malachite leaves a telltale green stain, indicating that there is some form of mineralization present. Since it is a carbonate, it is often associated with limestones, which are calcium carbonate. In a pure form, malachite makes for a nice specimen and will take a high polish. The key characteristic for identifying malachite is its bright green appearance. Malachite also has a light green streak. Malachite-rich stalactites often display ornate banding, offering many lapidary possibilities.

There are thousands of malachite locales across the United States. Search in known copper-producing districts.

Meteorite—Iron

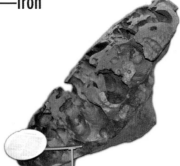

The massive Willamette meteorite was recovered near Portland, Oregon. It weighs over 30,000 pounds and resides at the American Museum of Natural History in New York City.

90 percent nickel to stony

Family: Extraterrestrial

Mohs: Varies by nickel-iron content

Specific gravity: Varies by nickel-iron content

Key test(s): Widmanstätten pattern

Likely locale(s): Anywhere, but deserts have slower oxidation

Most meteorites are actually stony, or primarily stone, but have enough iron to attract a magnet on a string, so the magnet test still works. Nickel is also a key component, measuring up to 7 percent, and many metal detectors can indicate the presence of nickel. After the magnet test, and barring a metal detector, look for thumbprint-shaped divots called regmaglypts, which are evidence of plenty of heat generated while passing through the atmosphere. Meteorites leave no streak, so a specimen with a reddish brown streak test is most likely hematite; a blackish-gray streak is magnetite. Scientists can slice these rocks with a saw and use a strong acid to etch the polished surface, resulting in a characteristic X-shaped structure known as a Widmanstätten pattern.

The best place to scout for meteorites is in a known "strewn field" surrounding an observed fall. Barring that, desert environments are productive because the limited moisture ensures that the iron-rich rocks do not quickly rust and fall apart. Many of the flat deserts in Basin and Range country offer meteorite-hunting possibilities.

MINERALS

Meteorite—Stony

This H5 chondrite is from Plainview, Texas, and features a dark, black fusion crust.
PHOTO COURTESY OF THE RICE NORTHWEST MUSEUM OF ROCKS AND MINERALS

90 percent nickel to stony

Family: Extraterrestrial

Mohs: Varies by nickel-iron content

Specific gravity: Varies by nickel-iron content

Key test(s): Blackened crust

Likely locale(s): Anywhere, but deserts have slower oxidation

Most meteorites are stony, or primarily stone, but with enough iron to attract a magnet on a string. The top clues between meteorites and meteor-wrongs rely on sight. For example, look for the presence of a fusion crust, which is simply evidence that the rock burned its way through Earth's atmosphere. This will typically appear as a black skin; if the rock has been broken or fractured, look for a bright interior beneath the black crust. The presence of chondrules, which are rounded grains, is another clue.

As mentioned for iron meteorites, the best place to scout for stony meteorites is in a known "strewn field" surrounding an observed fall. Desert environments are productive because the iron-rich rocks don't rust and quickly fall apart.

Molybdenite

Rare molybdenite crystal from Washington
PHOTO COURTESY OF THE RICE NORTHWEST MUSEUM OF ROCKS AND MINERALS

Molybdenum sulfide, MoS_2

Family: Metal sulfide

Mohs: 1–1½

Specific gravity: 4.6–5.1

Key test(s): Green streak; soft and greasy

Likely locale(s): Interesting mineralized zones

Molybdenite is the chief ore of molybdenum. It is somewhat rare and usually occurs only as small, dull-gray lumps. It has a metallic luster that could be confused with galena, and molybdenite is dark blue-gray to lead gray. One interesting field test is that molybdenite creates a green streak when scraped on a streak plate. When big enough to conduct a scratch test, molybdenite is extremely soft; a fingernail should scratch it easily. Crystals are rare, hexagonal, and tabular, but easily deformed. Fracture is unlikely, as molybdenite is so flexible and soft. Graphite is also lead gray but is lighter than molybdenite. Galena is also gray and soft, but it has a blue-gray streak and cubic crystal structure.

Many western states host molybdenite specimens, but collectible crystals are rare.

Platinum

Platinum flakes from Olivine Creek on the Simalkameen River, showing dark, silvery appearance

Pt
Family: Elemental metal
Mohs: 4–4½
Specific gravity: 21.5
Key test(s): Silvery gray; heavy
Likely locale(s): Black sands; accessory in metals mining

Platinum is a dull white-gray metal that looks somewhat like silver but is at least twice as dense. Platinum is also far more rare than silver, occurring mostly as an accessory in black sands with other Platinum Group Metals (PGMs), which include ruthenium, rhodium, palladium, osmium, iridium, and platinum. Because pure platinum does not oxidize, it is useful in jewelry; it has industrial uses as well. Because it alloys with other metals that do oxidize, flakes and nuggets in the field often appear dark gray or black.

Most of the world's platinum comes from South Africa. Many beach placers in Oregon, Washington, and California contain traces of platinum and other rare elements. Platinum deposits are widely scattered across the West, usually where there are significant chromite mines. The Stillwater Complex, a magnificent layered intrusive, produces the bulk of the platinum in Montana. In Colorado the mines at Breckenridge, Telluride, Ouray, and Aspen have all contributed platinum. California's Trinity River carries platinum values associated with chrome deposits in the hills.

Pyrite

Brassy pyrite crystals from Colorado
PHOTO COURTESY OF THE RICE NORTHWEST MUSEUM OF ROCKS AND MINERALS

Iron sulfide, FeS$_2$
Family: Metal sulfide
Mohs: 6–6½
Specific gravity: 4.9–5.1
Key test(s): Hardness
Likely locale(s): Known sulfide areas

Pyrite, or iron pyrite, is also known as "fool's gold," thanks to its brassy, yellowish color. Pyrite generally forms in the cubic crystal habit and often has striations, or lines, on its crystal faces. Other times, pyrite crystals tend to twin, interlock, and form interesting masses. There are dozens of varieties of pyrite, with various replacements for the iron (Fe) ion. Pyrite tarnishes rapidly, becoming darker and somewhat iridescent as oxygen attacks the iron-sulfur bond. Pyrite leaves a brownish-black or even greenish-black streak that smells faintly of sulfur, depending on how big a streak you make. Pyrite is hard, registering as high as 6½ on the Mohs scale, just below quartz. Many forms of pyrite are highly collectible.

Pyrite is found in every metal-mining district where sulfides are common.

Rutile

Fine rutile specimen from Graves Mountain, Georgia
PHOTO COURTESY OF THE RICE NORTHWEST MUSEUM OF ROCKS AND MINERALS

Titanium dioxide, TiO$_2$
Family: Metal oxide
Mohs: 6–6½
Specific gravity: 4.2
Key test(s): Luster
Likely locale(s): Pegmatites, skarns; granite

Rutile is usually dark, small, and opaque; but in pegmatites it can form nice crystals, usually as stubby pyramids. Color can vary from red, blood red, to brownish yellow, even violet, so that's not a good test. Rutile has an adamantine luster, and some crystal faces may show striations or lines. Rutile can concentrate in placer concentrates and dark beach sands, sometimes in economic quantities. Streak is not a good test for rutile; it is sometimes reported as bright red, sometimes as black, and sometimes as brown or light yellow. Interestingly, one of the primary uses for titanium other than in metallurgy is as a white pigment, especially in sunscreen. Rutile often forms as thin needles in garnet, sapphires, rubies, and quartz, resulting in asterism, or a star pattern.

Rutile is common in heavily metamorphosed rocks and in igneous rocks across the United States. Notable specimens come from Graves Mountain in Georgia and Southern California, but many states host rutile deposits.

Scheelite

Orange chunk of crystalline scheelite
PHOTO COURTESY OF THE RICE NORTHWEST MUSEUM OF ROCKS AND MINERALS

$CaWO_4$
Family: Tungstate
Mohs: 4½–5
Specific gravity: 5.9–6.1
Key test(s): Silvery (when fresh); tarnishes quickly
Likely locale(s): Economic ore deposits

Scheelite is the most common ore for tungsten, a metal important in the steel industry. Scheelite is typically yellow to yellow-orange but can be white, colorless, light gray, or other pale shades. Pure scheelite fluoresces a nice sky blue under shortwave ultraviolet light. Scheelite has a vitreous luster, has good cleavage, and forms excellent crystals that imitate diamonds. Scheelite can occur in granite pegmatites but is most common in hydrothermal veins where tin or gold is also present.

Scheelite is another metal sulfide common to the mining districts of the United States.

Siderite

Siderite in a yellow-brown mass
PHOTO COURTESY OF THE RICE NORTHWEST MUSEUM OF ROCKS AND MINERALS

Iron carbonate, $FeCO_3$
Family: Carbonates
Mohs: 3½–4½
Specific gravity: 4
Key test(s): Yellow-brown coatings
Likely locale(s): Bedded sedimentary deposits

Siderite is typically yellow, brown, or tan. Siderite crystals are rare; instead, siderite usually occurs as yellow-brown masses. It often occurs in hydrothermal veins and in bedded sedimentary deposits rich in calcium carbonate, but siderite also occurs as concretions in shales and sandstones. It is rich in iron by weight and is an important iron ore because it contains little sulfur. Siderite crystals are most common in pegmatites, occurring as trigonal or hexagonal tabs with vitreous, silky, or pearly luster. Siderite has a white streak.

There are numerous siderite locales across the United States, with Idaho and Colorado producing notable specimens.

Silver

Native silver; note the pinkish tint of "horn silver" in the sliced sample, lower left.

Ag
Family: Elemental metal
Mohs: 2½–3
Specific gravity: 10.1–11.1
Key test(s): Silvery (when fresh); tarnishes quickly
Likely locale(s): Economic ore deposits

Silver is actually rare in its native state because it forms oxides so readily. Isometric crystals are especially rare and very collectible. Silver has a characteristic shiny, metallic luster when fresh, appearing light gray or whitish gray at times. However, silver tarnishes quickly to black, brown, or yellow. Two key tests are for hardness and specific gravity. Lead forms crystals easier and has the characteristic stair-step pattern. Platinum is heavier than silver and even more rare, but some regions are noted for platinum nuggets, so do your research. Silver specimens almost always have some gold present, forming what the miners called "electrum."

Most western states have notable silver-mining districts. Montana, Idaho, Colorado, and Nevada all supply important specimens.

Sphalerite

Large sphalerite crystal on a bed of chalcopyrite and dolomite
PHOTO COURTESY OF THE RICE NORTHWEST MUSEUM OF ROCKS AND MINERALS

(Zn,Fe)S
Family: Metal sulfide
Mohs: 3½–4
Specific gravity: 3.9–4.2
Key test(s): Pale-yellow or light brown streak
Likely locale(s): Economic ore deposits

Sphalerite is the primary ore for zinc and is often very dark, depending on how much iron is present. It can also be green, yellowish, or red if the iron content is low. Well-formed crystals tend to be resinous or greasy black and are rare. Some varieties fluoresce. Most common forms of sphalerite present as heavy, dark masses, often with a reddish tint.

Sphalerite makes up much of the zinc ore in Idaho's Silver Valley. Most western states contain sphalerite locales in the usual sulfide-rich mining districts for those states.

Stibnite

Attractive blades of stibnite in crystalline form
PHOTO COURTESY OF THE RICE NORTHWEST MUSEUM OF ROCKS AND MINERALS

Antimony sulfide, Sb_2S_3
Family: Metal sulfide
Mohs: 2
Specific gravity: 4.6
Key test(s): Soft, slender, gray crystals
Likely locale(s): Hydrothermal deposits

Stibnite is the principal ore of antimony, and it is highly toxic. It has a metallic-gray luster and streak and is exceedingly soft, so it is easy to crush into a powder. Known as "kohl" in the ancient Middle East, it served as a cosmetic eyeliner, which would not have been healthy. Stibnite forms long, slender crystals, some quite stunning, and makes an attractive addition to your collection. It is often associated with arsenic minerals such as orpiment, realgar, and arsenopyrite in metal-rich hydrothermal vein systems.

Large deposits of stibnite are rare. The antimony mines on Stibnite Hill in Montana's Sanders County produce good specimens. Idaho has dozens of prospective sites and mines, especially around the appropriately named town of Stibnite.

Tetrahedrite

Attractive blades of tetrahedrite show beautiful crystallization.
PHOTO COURTESY OF THE RICE NORTHWEST MUSEUM OF ROCKS AND MINERALS

$(Cu,Fe)_{12}Sb_4S_{13}$
Family: Metal sulfide
Mohs: 3½–4
Specific gravity: 4.97
Key test(s): Distinctive crystals
Likely locale(s): Base metal zones

This is another mineral that represents a series, being the antimony end of a changing formula with the arsenic mineral tennantite at the other end of the scale. Another variation, freibergite, can be quite rich in silver. In its pure form, tetrahedrite forms unmistakable tetrahedrite crystals, hence the name, with distinct triangles sometimes flattened at each point. Tetrahedrite commonly occurs in massive form, being dark gray before rusting and turning reddish or yellowish.

Most western states host interesting tetrahedrite locales, including the famed Silver Valley, where it was a major ore. Montana also contains dozens of good locales, with good specimens from Silver Bow County and the Philipsburg district. Colorado is loaded with tetrahedrite sources, with excellent crystals coming from the famed Sweet Home Mine in the Alma district.

Ammolite

Raw ammolite tends to fracture and can be difficult for lapidarists to work with.
RAW AMMOLITE PHOTO FROM THE DISPLAY OF THE GEMOLOGICAL INSTITUTE OF AMERICA

$CaCO_3$
Family: Carbonates
Mohs: 4½–5½
Specific gravity: Variable
Key test(s): Iridescence
Likely locale(s): Ammonite fossil locales

Ammolite is a vivid, multicolored variety of ammonite fossil that has not only preserved but enhanced the nacre (NAY-ker), or colored shell. If you've ever seen an abalone shell, for example, you know the beautiful colors that display. Ammolite, fossilized ammonite shell, is highly prized by jewelers for its iridescent play of colors. Striking orange, green, red, yellow, and blue colors all vie for attention in ammolite. Since ammolite is often very thin and quite fragile, jewelers and lapidarists layer it with harder material for protection.

Most of the world's jewelry-grade ammolite originates in Canada. However, equivalent rocks to Canada's Cretaceous Bearpaw Formation extend into Colorado, with decent production.

Aquamarine

Aquamarine can form as short, stubby crystals or as longer, clear "pencils."
PHOTO COURTESY OF THE RICE NORTHWEST MUSEUM OF ROCKS AND MINERALS

$Be_3Al_2(SiO_3)_6$
Family: Beryl
Mohs: 7½–8
Specific gravity: 2.8
Key test(s): Hardness
Likely locale(s): Granite pegmatites

Aquamarine is another gem variety of beryl, like emerald, heliodor, morganite, red beryl, and goshenite. Aquamarine is blue, and when deep, striking blue, it is highly sought as a gemstone. It is quite rare, thus worth prospecting for. The hardness of around 8 makes it difficult to create a streak, but it is white. The luster is vitreous to resinous, cleavage is poor, and it does not fluoresce. It is usually associated with granite pegmatites.

Colorado's Mount Antero is the most famous aquamarine producer in the United States. The Pala Mine in San Diego County, California, is another noted locale. Idaho, New Mexico, New England, New Jersey, Virginia, and Georgia host aquamarine deposits; North Carolina boasts multiple locales.

Diamond

Loose diamond showing tetrahedral shape
PHOTO COURTESY OF THE RICE NORTHWEST MUSEUM OF ROCKS AND MINERALS

Carbon, C
Family: Native minerals
Mohs: 10
Specific gravity: 3.5
Key test(s): Hardness
Likely locale(s): Associated strictly with kimberlite pipes

Diamond specimens are clear, yellow, pink, blue, purple, brown, and even black. The hardness test is the best indicator—pure, strongly crystallized diamond, at 10 on the Mohs scale, scratches everything. Another good indicator is that diamond has an adamantine, or outstanding, luster. The crystal habit is octahedral, and cleavage is perfect in four directions, which is rare. Clear agate and clear gem-quality quartz such as Herkimer "diamonds" look similar to actual diamonds but are hexagonal and will not cleave.

Low-quality diamonds were reported from multiple placer operations throughout California, dating to the gold rush days. Placer camps from Wisconsin to North Carolina all recovered small diamonds in their concentrates. Many such gold mines throughout the West reported diamonds, although few samples remain. Crater of Diamonds State Park in Arkansas is the only fee-dig diamond operation in the United States. Diamonds are associated with kimberlite pipes along the Wyoming-Colorado border, and there has been considerable exploration; staking; and, for a limited time, meager production.

Emerald

Emeralds are difficult to work with, as they crack and fracture easily. Clarity is a key component for grading gems such as emeralds.

$Be_3Al_2(SiO_3)_6$
Family: Beryl
Mohs: 7½–8
Specific gravity: 2.8
Key test(s): Hardness
Likely locale(s): Granite pegmatites

Emeralds are a gem variety of beryl, associated with aquamarine, heliodor, morganite, red beryl, and goshenite. Emeralds tend to be cloudy and fractured, reducing their value, and they are quite rare in North America. The more brilliant greens are highly sought. The hardness of around 8 makes it difficult to take a streak, but it is white. The luster is vitreous to resinous, cleavage is poor, and it does not fluoresce. Emerald is usually associated with granite pegmatites.

There is a minor emerald exposure in Montana and another small occurrence in Connecticut. North Carolina is by far the biggest producer of emeralds in the United States; the Hiddenite Mine in Alexander County is a noted producer. British Columbia, Ontario, and the Yukon and Northwest Territories of Canada also contain emerald occurrences.

Garnet, Star

Polished triple star garnet from Emerald Creek, Idaho

Almandine, $X_3Y_2(SiO_4)_3$
Family: Metal silicates
Mohs: 6½–7½
Specific gravity: 3.6–4.3
Key test(s): Hackly fracture; hardness
Likely locale(s): Schist; black sands

Gem-quality garnet is somewhat rare; adding further value is a tendency for impurities to line up and create four- or six-sided stars. For instance, white rutile is the source of the star asterism at Emerald Creek. The USDA Forest Service operates a fee-dig operation at Emerald Creek, where the public can wash the gravels for these prizes. Polishing the garnets to reveal the prized stars requires patience, experience, and good equipment.

Several areas around Clarkia, Idaho, produce star garnet rough—check out *Rockhounding Idaho* (FalconGuides) for more information. Most other garnet locales in the United States do not yield gem-quality material unless found in pegmatites.

Jade

Jade specimens from Wyoming's jade fields

Manganese silicate, MnSiO$_3$

Family: Silicates

Mohs: 6½–7 for jadeite; 5½–6 for nephrite

Specific gravity: 2.9–3.1

Key test(s): Can't be scratched by a knife; botryoidal

Likely locale(s): Mafic rocks

There are two main varieties of jade. Both are amazingly strong due to interlocking crystals that form nearly unbreakable bonds. First, there is classic jadeite, a sodium-rich, aluminum-rich pyroxene found mostly in Burma and favored by Chinese rulers for centuries. Jadeite is not quite as hard as quartz. Second, there is nephrite jade, a type of amphibole, also found in China but more commonly associated with North American deposits in Wyoming, Washington, California, and British Columbia. Nephrite is slightly softer than jadeite and usually occurs only as white or shades of green. A high-quality steel knife blade cannot scratch either jadeite or nephrite. Jadeite, which can be purple, blue, lavender, pink, or vivid green, is the more highly prized, but that is not a fixed rule—rare "mutton-fat" white nephrite jade commands fabulous prices.

Wyoming's jade fields occupy a zone in the central part of the state near Jeffrey City, and there are multiple collecting locales described in *Rockhounding Wyoming* (FalconGuides). Nephrite jade and/or jadeite also is available in Washington, California, Alaska, and other states.

Precious Opal

This precious opal specimen is from Spencer, Idaho.
PHOTO COURTESY OF THE RICE NORTHWEST MUSEUM OF ROCKS AND MINERALS

$SiO_2 \cdot nH_2O$

Family: Metal silicates

Mohs: 5½–6½

Specific gravity: 2.0–2.2

Key test(s): Play of color; luster

Likely locale(s): Tuff deposits

Precious opal is a rare form of common opal that owes its stunning color to the way its silica spheres are stacked and packed, diffracting the light and causing the color interplay. Precious opal is relatively soft; has no crystal structure and no cleavage; and usually occurs in veins, coatings, and nodules. The streak is white. Precious opal comes in a variety of forms: standard precious opal, displaying all colors of the rainbow, usually after replacing wood; fire opal, usually red or orange, derived from thin seams between lava flows; black opal, very dark in color but with a nice play of colors; and white opal, which is usually white but also exhibits a full play of colors.

The Spencer Opal Mine, a fee-dig operation located near I-15 by Spencer, Idaho, boasts world-class precious opal. The Virgin Valley logs in northern Nevada are also considered exceptional and well worth a visit. Cedar Rim in central Wyoming also hosts abundant opal deposits. Fire opal, another sought-after form of precious opal, is found in various western states.

Sapphire

Variety of rough and faceted sapphires, including star sapphire

Aluminum oxide, Al_2O_3
Family: Oxides
Mohs: 9
Specific gravity: 3.9–4.0
Key test(s): Hardness
Likely locale(s): Pegmatites, dikes, and metamorphic zones

Sapphire and ruby are both varieties of gem-quality corundum. Unlike rubies, which are typically red, sapphires are usually some form of blue, although sapphires can be clear, light gray, or even dark gray. The blue color comes from iron and titanium impurities that affect ion charges, color absorption—it all gets very technical. The inclusion of rutile needles results in a "star" effect for properly polished sapphires and is highly desirable. Note that sapphires are one of the easiest gems to treat with heat to enhance or change the color. Sapphires are typically associated with pegmatites and can be difficult to separate from host rock.

Sapphire is associated with pegmatites in California, Idaho, Montana, Colorado, and North Carolina. The Spokane Bar area near Helena, Montana, hosts a popular fee-dig operation. Franklin County, North Carolina, also contains a fee-dig operation. The distinctive blue Yogo sapphire is one of the better-known US gems.

Topaz

Large rough topaz will scratch quartz, and its luster is different from that of quartz, calcite, or other similar minerals.

PHOTO COURTESY OF THE RICE NORTHWEST MUSEUM OF ROCKS AND MINERALS

Aluminum silicate, $Al_2SiO_4(F,OH)_2$

Mohs: 8

Specific gravity: 3.4–3.6

Key test(s): Pink (when fresh); hardness; glassy or vitreous luster

Likely locale(s): Rhyolites

Topaz is an interesting mineral that does not seem to get much respect for its natural state. Gem traders sometimes irradiate topaz to produce results that are more pleasing. Typically, topaz is clear when pure, but impurities can cause it to appear white, light gray, or even pink or yellow. Laboratory treatments result in more stunning blues and orange colors. Topaz has a glassy to vitreous luster and forms stubby, prismatic crystals in the orthorhombic crystal system, with excellent terminations. The best test in the field is its hardness—it will scratch quartz. Cleavage is excellent in one direction, and striations are common lengthwise on crystal faces, but topaz often occurs in massive lumps. Topaz forms at high temperatures, among silica-rich igneous rocks such as rhyolite or in cavities within granite pegmatites.

Topaz occurs in Colorado at Lake George and elsewhere in that state. Idaho and Montana report topaz occurrences; Arizona, California, Nevada, and New Mexico also feature this gem. There are multiple locales across the eastern United States, from Alabama to New England; Wisconsin, Michigan, and Missouri have also reported topaz finds. Utah's Topaz Mountain is a famed digging area in rhyolite.

Tourmaline—Elbaite

"Watermelon" tourmaline (elbaite) from the Himalaya Mine, Southern California
PHOTO COURTESY OF THE RICE NORTHWEST MUSEUM OF ROCKS AND MINERALS

$Na(Mg,Fe)_3Al_6(BO_3)_3(Si_6O_{18})(OH,F)_4$
Family: Cyclosilicates
Mohs: 7–7½
Specific gravity: 3.0–3.3
Key test(s): Color bands; striations; harder than apatite
Likely locale(s): Pegmatites

Tourmaline comes in a variety of forms, with the most common being black schorl (see page 103). Semiprecious gem varieties include elbaite, first noted on Elba Island but now primarily known from Southern California; indicolite, which is blue; rubellite, which can be pink or red; and dravite, which is brown. Tourmaline has a vitreous luster, leaves a white streak, and forms a hexagonal crystal with no cleavage. Hornblende is also black, but schorl has a triangular cross section. Gem-quality tourmalines are usually found in vugs and cavities within granite pegmatites, such as in Maine and Southern California.

Turquoise

Polished turquoise can be pure creamy blue or contain black veins and inclusions.
PHOTO COURTESY OF THE RICE NORTHWEST MUSEUM OF ROCKS AND MINERALS

$CuAl_6(PO_4)_4(OH)_8 \cdot 4-5H_2O$

Family: Phosphates

Mohs: 5–7

Specific gravity: 3.5–4.0

Key test(s): Blue; waxy luster

Likely locale(s): Veins, seam fillings; near copper mines

Natural turquoise rarely forms crystals, occurring instead as nuggets and filled-in fractures in the host rock. Turquoise frequently forms veins and nodules as a secondary replacement mineral. Turquoise has a waxy luster and a faint, bluish-white streak; the powder is soluble in hydrochloric acid. Under long-wave UV light, turquoise may fluoresce. Black limonite veining is common. Minerals sometimes confused with turquoise include chrysocolla, which is much softer, and variscite, which is usually greener as well as softer. Look for turquoise in known copper-producing areas where phosphates are also common.

Turquoise is often associated with silver jewelry from Arizona, but California, Nevada, Utah, Colorado, and New Mexico also host turquoise locales. East of the Mississippi River, turquoise is reported from Arkansas, Alabama, Virginia, Pennsylvania, and other locales.

Acknowledgments

Special thanks to the staff at the Rice Northwest Museum of Rocks and Minerals in Hillsboro, Oregon. I'd like to single out Julian Gray and Leslie Moclock for their assistance.

Special thanks to my family and especially my wife, Cindy, who has supported me over all these years. Also thanks to Rachel Houghton, veteran technical communicator and longtime friend, who helped with photography, touch-up, editing, and encouragement, and to Martin Schippers, Frank Higgins, and Dirk Williams, who have been my most faithful assistants in the field.

Rice Northwest Museum of Rocks and Minerals (ricenorthwestmuseum.org)

Glossary

alluvium: Dirt, usually. Stream and river deposits of sand, mud, rock, and other material. Sometimes sorted, if laid down in deep water; otherwise can be unsorted if deposited during floods, earthquakes, etc. If glaciers were involved, the term *till* is used.

anthracite: The hardest and most intensely metamorphosed form of coal.

arkose: Sandstone that has many unsorted, broken-up pieces of feldspar and quartz. Usually hard and not easily eroded.

basement: The "lowest" and oldest rocks around, usually metamorphic and frequently dating to the Precambrian or Paleozoic age. Basement rocks are usually less prone to erosion and make up mountain ranges and stunning cliffs. "Basement" refers to their placement at the bottom of a stratigraphic table.

batholith: General term that refers to extremely large masses of coarse intrusive rock such as granite. Any rock formation over 100 km² (39 square miles) is considered a batholith.

bedding: The tendency of sedimentary rocks such as sandstone to reside in visible zones or marker beds, similar to tree rings.

bleb: A round or oval cavity, air bubble, hole, or vesicle, usually in basalt, and sometimes filled with opal, agate, or chalcedony.

chemical sediment: Refers to the way certain limestones and dolomites precipitate material such as calcium carbonate, which falls to the bottom of the sea or bay and accumulates.

clasts: Catchall term for the clay, silt, sand, gravel, cobbles, and boulders that make up nonchemical sedimentary rocks. The size of the clasts then determines the name of the rock.

clay: Usually refers to the smallest mineral fragments, smaller than 2µm or $\frac{1}{255}$ millimeter.

cobble: Fancy term for rocks between a pebble and a boulder. The exact definition of a cobble is anything from 64 to 256 millimeters in size.

contact metamorphism: The result of a hot igneous intrusion on the country rock. The contact zones between the intrusion and the surrounding rock can sometimes house interesting mineralization.

density: Defines the weight per an agreed unit of volume. By weighing the sample and then dunking it in water and measuring the volume of water displaced, we get the density measured in grams per cubic centimeter. The general term *heft* refers to a field test for how dense a typical-size hand specimen feels.

diatomite: Usually a white, chalky deposit that, upon microscopic inspection, turns out to be composed of tiny diatom fossils. These beds can often host common opal, precious opal, and zeolite deposits.

drift: General term for glacial deposits composed of jumbled debris. Outwash plains and terraces are usually sorted, whereas tills and moraines are unsorted.

dry wash: The sign of a seasonal stream that dries up during the summer months. These can be interesting for rockhounds—specimen sizes are usually bigger because they have not been severely eroded.

eolian: General term for wind deposits such as loess, sand sheets, ripples, and dunes; also can refer to wind processes such as dust storms, sandblasting, and desert varnish.

eon: The longest division of geologic time is the super-eon. The model is super-eon ⟶ eon ⟶ era ⟶ period ⟶ epoch ⟶ age. Thus, we are in the Holocene epoch of the Quaternary period of the Cenozoic era of the Phanerozoic eon of the Cambrian super-eon.

epoch: Shorter subdivision of a geologic period, usually corresponding to observed stratigraphy in the field.

era: The four main geologic eras, from oldest to youngest, are the Precambrian, Paleozoic (which starts with the Cambrian), Mesozoic (the "age of reptiles"), and Cenozoic (ours).

erosion: The forces and processes that continually grind down mountains and move their debris downwind or downhill.

evaporite: As bodies of water dry up under desert conditions, they frequently get white or light brown crystalline rings around the edges. These minerals are usually pure salt (halite, or sodium chloride) or a related halide, plus borax or gypsum as well.

exfoliation: Refers to the way rocks, and in particular granite, tend to slough off skins or layers of outer rock, like an onion. The result is usually a rough, rounded shape rather than angles and edges.

facies: Field term used to describe how sedimentary rocks can be identified by the way they were deposited. The term *biofacies* describes the distinct fossil assemblage, while the term *lithofacies* could describe the similarities in a rock's clast size. There could be several distinct facies identifiable in the field that make up an overall formation.

float: Describes the difference between rock samples hammered from an outcrop, and thus with a known origin, and samples that exist as cobbles or boulders and not attached to bedrock. Prospectors are able to trace float to its source outcrop.

flood basalt: Refers to the way basalt tends to pour out of cracks and vents and form rivers of liquid rock, thus creating plateaus of flat, layered deposits. By comparison, andesites pile up.

flow cleavage: Describes the tendency of metamorphic rocks to arrange flat, elongated crystals into a parallel structure.

formation: This is a key term to understand in field geology. Geologists assign formation names to mappable, recognizable rock assemblages and also note a "type" locale that defines the rest of the unit. Formations can be lumped together into groups or even supergroups. To be considered a formation, a group of similar rocks must be big enough to be worth the bother, be the same age, share some key similarity, and be traceable across the surface. Some formations are divided into members.

geologic cycle: The continuous cycle of destruction, recycling, and rebirth that defines the way Earth's crust works. There are countless variations on the scenario, but in general, rocks are created, such as by a volcano; eroded; built back into sedimentary deposits; subducted; cooked; melted; then turned back into lava and erupted again.

graded bedding: Sedimentary rock term to describe the way a creek or river deposits alluvium, pebbles, and boulders in a typical sequence, from coarse to fine. Look for coarse conglomerates at the bottom and fine siltstones at the top.

gravel: The best place for pebble pups and rockhounds to search for interesting material. Gravels usually consist of pebbles, cobbles, and boulders, in various ratios, and also contain varying amounts of sand and silt. Gravel bars refresh with each season and provide clues to the surrounding geology.

hydrothermal vein: Best spot to investigate for interesting minerals. These hot, chemical-rich solutions, usually quartz, can either find an existing crack in country rock or create their own. If they cool slowly enough, hydrothermal veins can create ore deposits with large crystals that are prized by collectors.

ignimbrite: The igneous rock created when hot volcanic ash and breccias pour out of a volcano or vent and are too heavy to drift away as an ash cloud.

intrusion: Catchall term for the various granites, diorites, and related rocks that bulldoze their way through Earth's crust but never reach the surface. Cooling in place quickly results in fine-grained material; cooling slowly gives the individual elements more time to build up into larger crystals.

laccolith: A small intrusion that squeezes in horizontally between beds and builds itself out laterally. These deposits usually have a neck, where material fed in, and a dome, depending on how forceful the intrusion was.

lahar: When volcanic eruptions mix with melted glaciers, lakes, and snow, the result is a dangerous mudflow called a lahar. The material can flow quickly, but it will soon solidify into a concrete-hard mass of jumbled-up ash, fragments, and pumice.

lava: Catchall term for extruded molten rock that includes basalt, andesite, rhyolite, dacite, and other materials.

lode: The prized zone of rich, extended mineralization that usually ensures a successful mining operation. The term is reserved for larger vein networks that cover significant ground.

luster: Term for the visual appearance of a mineral's lighted surface. The way minerals reflect light can be helpful for identification, but terms such as *metallic* and *waxy* are not completely standardized.

mafic mineral: Dark, heavy minerals that are rich in iron and magnesium, such as pyroxenes, amphiboles, and olivines.

magma: The molten lava that eventually forms igneous rock when it cools. Magma that cools without eruption is called an intrusion; otherwise it is extrusive.

magma chamber: The source cavity or reservoir for magma traveling up through Earth's crust.

mass spectrometer: The one instrument you wish you had, because it can count ions and provide their exact distribution. An inexpensive, handheld mass spectrometer would revolutionize field geology.

Mohs scale of hardness: The observation-based method of ranking a mineral by what it can scratch and what, in turn, can scratch it. Diamond is alone at the top of the list, at 10; it can scratch corundum, which can scratch topaz, which can scratch quartz, at 7. The remaining minerals on the scale are orthoclase feldspar, 6; apatite, 5; fluorite, 4; calcite, 3; gypsum, 2; and talc, the softest, 1.

native metal: Metals in their purest form are not significantly combined with oxides, sulfides, carbonates, or silicates and are thus native. Gold, silver, copper, platinum, and mercury are examples.

oil shale: A dark, organic-rich shale that sometimes contains enough petroleum-based ingredients to burn.

oolite: Refers to the small, round form taken when calcium carbonates start to coat sand grains and roll around in a lime-rich sea bottom.

ore: A viable mineral deposit that is worth mining. Usually refers to a metal-based mineral that must be milled.

Original Horizontality, Principle of: The idea proposed by Nicholas Steno (1638–1686) that sedimentary rocks are laid down flat. Since many sandstone beds are currently tilted, this simple concept meant other forces were at work.

outcrop: A cliff, ledge, or other visible clue to the rock formations below. Exposed, mappable basement rock.

pegmatite: A key igneous rock, usually found as a vein or dike, with very large grains of mica, feldspar, and tourmaline, among other minerals. Pegmatites tend to form cavities and vugs where large crystals can accumulate without crowding into one another and damaging their crystal structure. Pegmatites sometimes host smoky quartz, beryl, topaz, aquamarine, and other exotic gems.

pelagic sediment: Catchall term for the fine sediments that slowly accumulate in deep marine environments. Rather than relying on silt or clastic debris, this material is predominantly derived from shells of microscopic organizations such as foraminifera. Accumulation rates are as slow as 0.1 centimeter per 1,000 years.

reaction series: Refers to the behavior of a cooling magma where some minerals form at high temperatures, and others as temperatures cool. These conditions are observable in a laboratory setting.

regolith: Another catchall term, this time describing the various rock fragments and erosional debris that lie on bedrock, including alluvium and clastics.

replacement deposit: Describes a particular type of ore deposit where hot, circulating solutions first dissolve a mineral to form a cavity, then fill the void with a new material.

sedimentary structure: Describes the various relics of a sedimentary rock's deposition, such as ripples, cracks, bedding, zones, and layers.

stratification: The tendency of sedimentary rocks to form in flat, parallel sequences that are mappable at the surface for considerable distances. Geologists identify patterns and unravel the sequence of events represented in the strata.

stratigraphic column: Stratigraphy is the science and study of sedimentary rock outcrops. Stratigraphers draw stratigraphic columns that pictorially represent the measured or inferred relations of rock outcrops, especially age. Metamorphic and igneous rocks occasionally show up at the bottom in a stratigraphic column, as it is primarily a tool to understand sedimentary rocks.

streak: Refers to the color of the powdered mineral dust left behind when a mineral is scraped across a streak plate. The color of this fine rock powder is a more true reflection of the mineral than visual appearance. Since a streak plate is about 7 on the Mohs hardness scale, only minerals less than 7 can easily be tested by streak.

Superposition, Principle of: Another theory traced to Nicolas Steno (1638–1686), who pointed out that a rock formation or stratum that sits on top of another distinct layer must usually be younger than the rock below. The older rock will be at the bottom, except in rare conditions.

talus: The impressive accumulation of debris below a cliff or prominent outcrop. Because the rocks are shifting constantly and continuing to accumulate, few plants can get a foothold. Also called scree.

tectonics: Geologic theories of how Earth's crust continually moves and shapes new rocks. To compare planets, on Mars there is little evidence of tectonics, and there is basically one volcano, Olympus Mons, rising 13 miles (21 km) above the planet's surface. On Earth the plates continually shift, resulting in arcs, troughs, and collision zones.

texture: Describes a rock's grain size, crystal size, whether the grains are uniform or variable, whether the grains are rounded or angular, and whether there is any evidence of orientation to the grains.

till: Jumbled mess of glacial debris, with little to no recognizable bedding present and sediment sizes ranging from rock flour and clay all the way to massive boulders.

tuff: The term used to describe ash, pumice, volcanic breccias, and other debris. Some ash beds yield petrified wood and fossil leaves; others are welded solid.

ultramafic rock: Igneous rocks, such as dunite, peridotite, amphibolite, and pyroxenite, that consist of primarily mafic minerals and contain less than 10 percent feldspar.

volcanic ash: The fine rock fragments and glassy, angular material ejected from a volcano into the air.

volcanic breccia: A pyroclastic rock made up of angular fragments that show little or no sign of waterborne movement. Particle sizes are greater than 2 millimeters in diameter.

xenolith: Literally, "foreign rock" where a piece of country rock is picked off the walls or otherwise incorporated into a rising or spreading dike or intrusion.

zeolite: Common aluminosilicates formed in volcanic rocks, such as basalt, where alkaline groundwater circulates at low temperatures and creates a ringed "molecular sieve" structure.

Index

About the Author

Garret Romaine, an avid rockhound, fossil collector, and gem hunter, is the author of multiple field guides, handbooks, and identification guides. He was a columnist for *Gold Prospectors* magazine for fifteen years and is a member of the board of directors of the Rice Northwest Museum of Rocks and Minerals in Hillsboro, Oregon.